TESLA

THE
LIFE and TIMES OF AN
ELECTRIC MESSIAH

NIGEL CAWTHORNE

CHARTWELL
BOOKS

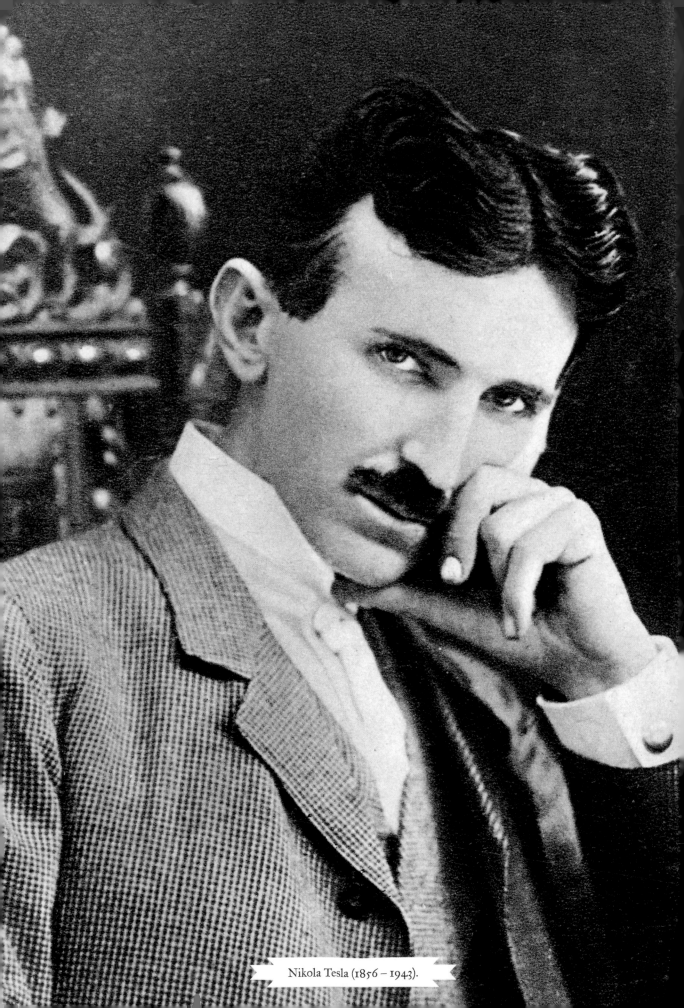

Nikola Tesla (1856 – 1943).

CONTENTS

If you mean the man who really invented, in other words, originated and discovered – not merely improved what had already been invented by others, then without a shade of doubt, Nikola Tesla is the world's greatest inventor, not only at present, but in all history.

Hugo Gernsback, Father of Modern Science Fiction, 1919[1]

Nikola Tesla (1856 - 1943) is the towering genius who made the modern world. All the electrical devices around us owe something to him. Not only did he invent many of the gadgets we depend on today, he had a vision of the future, much of which has become reality long after his death. As long ago as 1900, Tesla wrote of a world system of wireless transmission:

> *The World-System has resulted from a combination of several original discoveries made by the inventor in the course of long continued research and experimentation. It makes possible not only the instantaneous and precise wireless transmission of any kind of signals, messages or characters, to all parts of the world, but also the inter-connection of the existing telegraph, telephone, and other signal stations without any change in their present equipment. By its means, for instance, a telephone subscriber here may call up and talk to any other subscriber on the globe.*

This is surely the mobile phone network we have over a century later. And in his autobiography, *My Inventions*, published in 1919, he envisaged that in nine months, without undue expense, he could deliver:

✢ The interconnection of existing telegraph exchanges or offices all over the world;

✢ The establishment of a secret and non-interferable government telegraph service;

✢ The interconnection of all present telephone exchanges or offices around the globe;

✢ The universal distribution of general news by telegraph or telephone, in conjunction with the press;

✢ The establishment of such a 'World System' of intelligence transmission for exclusive private use;

✢ The interconnection and operation of all stock tickers of the world;

✢ The establishment of a world system of musical distribution, etc.;

✢ The universal registration of time by cheap clocks indicating the hour with astronomical precision and requiring no attention whatever;

✢ The world transmission of typed or handwritten characters, letters, checks, etc.;

+ The establishment of a universal marine service enabling the navigators of all ships to steer perfectly without compass, to determine the exact location, hour and speed; to prevent collisions and disasters, etc.;

+ The inauguration of a system of world printing on land and sea;

+ The world reproduction of photographic pictures and all kinds of drawings or records...[2]

Here we have the internet, GPS and Satnav. But Tesla was not just a visionary who delivered theory. He was a practical man who pioneered alternating current that made it possible to transmit electricity over long distances, allowing electrical appliances to be powered by remote power stations, rather than have a power station on every street corner as the earlier direct current system envisaged.

He is now acknowledged to have beaten Guglielmo Marconi to the invention of the radio. Indeed, he spoke of his world system of wireless transmission the year before Marconi transmitted the first radio signal across the Atlantic. His Tesla Coil, invented in 1891, is widely used in radio and television sets, and other electronic equipment. He developed electric motors, generators, X-rays, fluorescent tubes, remote control and radar. However, many of his inventions are unacknowledged because he was so busy developing new ideas to bother patenting them.

Although he was never awarded a Nobel Prize, three Nobel laureates lauded him as 'one of the outstanding intellects of the world who paved the way for many of the technological developments of modern times'.[3] He appeared on the cover of *Time* magazine and the unit of magnetic induction, a minor planet, a crater on the moon, an award and an airport are named after him. He was also played by David Bowie in the 2006 movie *The Prestige*.

Otherwise, with his talk of death rays and communication from other planets, his image endures as that of the mad scientist. But he was far from mad. He was one of the outstanding figures of the 20th century, arguably more influential than Einstein, Stravinsky or Picasso, and 70 years after his death, he deserves to be better known.

Nikola Tesla examining some of his induction motors, 1898.

PART ONE

STRANGER FROM A STRANGE LAND

CHAPTER 1

BIRTH OF A VISIONARY

In my boyhood I suffered from a peculiar affliction due to the appearance of images, often accompanied by strong flashes of light, which marred the sight of real objects and interfered with my thought and action. They were pictures of things and scenes which I had really seen, never of those I imagined. When a word was spoken to me the image of the object it designated would present itself vividly to my vision and sometimes I was quite unable to distinguish whether what I saw was tangible or not. This caused me great discomfort and anxiety…

Nikola Tesla[1]

Legend has it that Nikola Tesla - the pioneer who brought electric light to nearly every home on the planet - was born in a dazzling electrical storm. Sadly, the meteorological records of the Balkans in the 19th century are not readily to hand. But it would have been a fitting debut for the man who made his own artificial lightning with thunder that could be heard 15 miles (24 km) away. In 1894, before a large gathering of people in Philadelphia, Tesla ran 250,000 volts through his body to demonstrate the safety of alternating current. An eyewitness to his experiments said that there was 'light flaming at every pore of his skin, from the tips of his fingers and from the end of every hair on his head'.[2]

On the night of his birth, hearing the thunder, according to family lore, the fearful midwife said: 'He'll be a child of the storm.' His mother responded: 'No, of light.'[3]

Born in the dark of the small Croatian village of Smiljan at midnight on 10 July 1856, Tesla was ethnically a Serb. His family had left Serbia, then under the Ottoman empire, for Catholic Croatia when it became part of the burgeoning Austrian Empire in the 1700s. Both his grandfathers had fought in Napoleon's Illyrian Army that aimed to kick out the Hapsburgs and the Ottomans, and unite the Balkans.[4]

Tesla's father, Milutin, was an Orthodox priest and wanted Nikola to follow in his footsteps. Milutin was a poet and political activist who wrote of a united Yugoslavia, which did not come about until 1929. His mother Djouka never learned to read, but could memorize epic Serbian poems and long passages of the Bible. Tesla attributed his phenomenal memory to her.[5]

ELECTRIC PET THEORY

It was the Tesla family's cat, Macak, that introduced the 3-year-old Tesla to electricity. 'As I stroked Macak's back, I saw a miracle that made me speechless with amazement,' he recalled. 'Macak's back was a sheet of light and my hand produced a shower of sparks loud enough to be heard all over the house.' The young Tesla asked his father what had caused the sparks. Milutin replied: 'Electricity, the same thing you see through the trees in a storm.' It led Nikola to think that nature was a cat and God was stroking it.

That night, as it grew dark, Nikola noticed that the cat was surrounded by a halo from the static electricity in its fur. In 1939, looking back on the experience, Tesla said: 'I cannot exaggerate the effect of this marvellous night on my childish imagination. Day after day I have asked myself "what is electricity?" and have found no answer. 80 years have gone by since that time and I still ask the same question, unable to answer it.'[6]

MOTHER OF INVENTION

Tesla inherited his flair for inventing from his mother who, he said, 'descended from one of the oldest families in the country and a line of inventors. Both her father and grandfather originated numerous implements for household, agricultural and other uses.'[27] But she stood out even among this remarkable family. He said:

My mother was an inventor of the first order and would, I believe, have achieved great things had she not been so remote from modern life and its multi-fold opportunities. She invented and constructed all kinds of tools and devices and wove the finest designs from thread which was spun by her. She even planted the seeds, raised the plants and separated the fibres herself.[8]

He also inherited his immense appetite for work from her. 'She worked indefatigably, from break of day till late at night,' he said, 'and most of the wearing apparel and furnishings of the home was the product of her hands. When she was past sixty, her fingers were still nimble enough to tie three knots in an eyelash.'[9] While famous for her embroidery, she also devised a mechanical eggbeater.

INFAMOUS FROG CATCHER

In his autobiography *My Inventions*, Tesla recalled his first invention. One of his playmates had a fishing rod and set out with friends to catch frogs. But Tesla was left out because he had had a quarrel with the boy. So he got hold of a piece of soft iron wire, hammered the end into a point between two stones, bent it into shape and attached it to a strong string. Then he cut a branch to make a rod and, armed with some bait, went to the brook.

However, he found that, while the frogs would not take his bait, they would bite on the bare hook. He kept this secret from the other boys who caught nothing, only telling the secret at Christmas, in the generous spirit of the season.

Next, he made an early attempt 'to harness the energies of nature to the service of man,'[10] he said. He attached a rotor to a spindle with a disc on the other end in an attempt to make a primitive helicopter. To power the device, he attached four June bugs.

'These creatures were remarkably efficient,' he said, 'for once they were started, they had no sense to stop and continued whirling for hours and hours, and the hotter it was, the harder they worked.'[11] But then another boy came along and ate the June bugs alive. After that, Tesla was never able to touch another insect.

He took clocks to pieces and discovered how difficult it was to put them back together again. He also made himself a wooden sword and, imagining himself a great Serbian warrior, he slashed at cornstalks, ruining the crop and earning himself a spanking from his mother.

TRAGIC DEATH OF DANE

Tesla possessed a powerful imagination, at that time, filled with superstition and religious images. They were almost tangible and often accompanied by flashes of light that obscured the real world. He did not think this peculiar as his elder brother Dane saw the same things.

The young Nikola was overshadowed by Dane who, he said, was 'gifted to an extraordinary degree'.[12] But, when Nikola was 5, Dane was thrown by the family's Arabian horse and died of his injuries.

'I witnessed the tragic scene and, although so many years have elapsed since,' he wrote in 1919, 'my visual impression of it has lost none of its force.'[13]

However, there was another version of this story. In it Dane was said to have died after falling down the cellar stairs. He suffered a head injury and in his delirium accused Nikola of pushing him. Nearly 70 years later he still recalled the night of Dane's death:

It was a dismal night with rain falling in torrents. My brother, … an intellectual giant, had died. My mother came to my room, took me in her arms and whispered almost inaudibly: 'Come and kiss Daniel.' I pressed my mouth against the ice-cold lips of my brother knowing only that something dreadful had happened. My mother put me again to bed and lingering a little said with tears streaming: 'God gave me one at midnight and at midnight he took away the other one.'

Following Dane's death, his parents recalled his achievements, making Nikola's seem dull by comparison. This undermined his confidence and he ran away, seeking refuge in an inaccessible mountain chapel that was only visited once a year, remaining there 'entombed for a night'.[14]

The tragedy of Dane's death never left him. For the rest of his life he would have nightmares about it. And after Dane's death, Tesla's waking fantasies became more real and he began to have out-of-body experiences.

CHILDHOOD TRAUMAS

There were other traumatic events in Tesla's childhood. After having a bath on a summer's day, his mother put him outside naked to dry in the sun where he was attacked by a goose that seized him by the navel with its beak and almost pulled it inside out. He once fell headlong into a huge vat of boiling milk, risked drowning swimming under a raft and found himself almost swept over a waterfall created by a nearby dam. As well as being lost, frozen and entombed, he claimed to have had 'hairbreadth escapes from mad dogs, hogs, and other wild animals'.[15]

Even ordinary things held hidden terrors. He developed a strange aversion to women's earrings. The sight of a pearl would almost give him a fit, though he was fascinated by crystals. He would not touch other people's hair 'except, perhaps, at the point of a revolver'. He would get a fever from looking at a peach and hated having camphor anywhere in the house. Dropping little squares of paper into a dish filled with liquid produced an awful taste in his mouth. Some of these strange quirks helped prepare him for the world of science. He would count his steps as he walked and calculate the volume of soup plates, coffee cups and pieces of food. Otherwise he did not enjoy his meals. Every repeated act had to be done a number of times that was divisible by three. If not, he would start over.[16]

UNSETTLING TOWN LIFE

Soon after Dane's death, Tesla's father was promoted and moved to an onion-domed church in the town of Gospic. There Nikola started school. His father had a well stocked library, but flew into a rage when he discovered Nikola reading at night. Fearing the boy's eyesight would be strained, he hid the candles. Undeterred the enterprising Nikola cast his own and sealed up any cracks in his room so the light could not be seen from the outside. Then he read until dawn.

Nikola missed the countryside and found himself ill-equipped for life in the town. 'In our new house I was but a prisoner,' he wrote, 'watching the strange people I saw through my window blinds. My bashfulness was such that I would rather have faced a roaring lion than one of the city dudes who strolled about.'[17]

Then this shy boy met with an incident 'the mere thought of which made my blood curdle like sour milk for years afterwards'.[18] Coming down from the church belfry one Sunday after ringing the bell, he stepped on the train of one of the town's grand dames which 'tore off with a ripping noise which sounded like a salvo of musketry fired by raw recruits'. His father was livid and slapped him on the cheek. This was the only corporal punishment he ever administered.[19]

DEVELOPING MIND OVER MATTER

Until the age of 8, Tesla admitted that his character was 'weak and vacillating'. Then he came upon an historical novel called *Abafi* - which means 'Son of Aba' - by Hungarian writer Miklós Jósika. In it, the young roué Olivér Abadir gradually mends his ways and becomes a national hero in

Transylvania's fight against the onslaught of the Hungarians, Turks and Austrians. Following his example, Tesla set about developing willpower. 'In a little while I conquered my weakness and felt a pleasure I never knew before - that of doing as I willed,' he said.

Following the incident of the torn train, Tesla had been ostracized in Gospic. Now he managed to redeem himself. The town had recently organized a fire department and was showing off its new fire engine. The entire populace turned out for the ceremony and speeches. With the hose at the ready, the order was given to start pumping, but not a drop of water came out. While the bigwigs tried in vain to locate the trouble, Tesla felt instinctively for the suction hose that ran down into the river. He found it collapsed. Plainly there was a blockage, so he waded into the river and unblocked it. Suddenly he was the hero of the day and found himself carried shoulder high.

CALCULUS, COILS AND TURBINES

At 10 years old, Tesla entered the local Real Gymnasium - the equivalent of a British prep school or an American junior high school. It had a well-equipped physics department.

'I was interested in electricity almost from the beginning of my educational career,' he said. 'I read all that I could find on the subject ... [and] experimented with batteries and induction coils.'[20]

He was also keen on waterwheels and turbines, and experimented designing a flying machine which, he realized later, could not work because it depended on perpetual motion. Then, after seeing a picture of Niagara Falls, he told his Uncle Josif that one day he would go to America and put a big wheel under the falls to harness its power.[21]

Finishing at the Real Gymnasium at the age of 14, Tesla fell ill. During his youth he

claimed that three times he was in such a bad way that he was 'given up by physicians'.[22] While he was recuperating, the local library sent all the books it had not catalogued for Nikola to read and classify. It was then, for the first time, he came across the works of Mark Twain, whom he would later befriend.[23]

When he recovered, his father sent him to Karlovac - also known as Karlstadt - to the Higher Real Gymnasium to prepare him for the seminary. Nikola's father was still determined that his son should follow him into the priesthood, a prospect which filled Tesla with dread. At the Higher Real Gymnasium, he showed early signs of genius, performing integral calculus in his head, leading his teachers to think he was cheating.

Again the Gymnasium at Karlovac had a good physics department. Tesla became fascinated by the Crookes radiometer they had there.[24] Invented by British scientist William Crookes, it consisted of four metal vanes, polished on one side, blackened on the other, mounted on a vertical pivot in a glass bulb. The mechanism spun when bright light fell on it. It was also in Karlovac, in 1870, he saw, for the first time, a steam train.

CONTRACTING CHOLERA AND RECUPERATING

When he had completed his studies at Karlovac, Tesla got a message from his father telling him to go into the mountains with a hunting party. This puzzled him as his father did not approve of hunting, so he ignored the message and returned to Gospic to find it in the grip of a cholera epidemic. That was why his father wanted him to stay away. Nikola soon came down with the disease and was confined to bed for nine months. When he was at death's door, his father tried to encourage him in the hope he would rally. Nikola seized the opportunity and said to his father: 'Perhaps I may get well if you will let me study engineering.' His father replied: 'You will go to the best technical institution in the world.'[25] After that he pulled through.

During his recuperation, he would take long walks in the forest. Along the way he conceived a way to send letters and parcels between continents via tubes under the oceans. Mail would be packed in balls that would be forced along the pipes by water at high pressure. However, he did not take into account the resistance to the flow of water and the system would not have worked.

He also thought up a scheme to speed up worldwide travel. A stationary ring would be erected high above the equator, with the world turning underneath it. People would travel up onto the ring, then wait for their destination to appear below. He conceded that it would be impossible to build such a ring, but these thought experiments prepared his mind for later work. He said:

I observed to my delight that I could visualize with the greatest facility. I needed no models, drawings or experiments. I could picture them all as real in my mind. Thus I have been led unconsciously to evolve what I consider a new method of materializing inventive concepts and ideas, which is radically opposite to the purely experimental and is in my opinion ever so much more expeditious and efficient.[26]

As good as his word, Tesla's father secured a scholarship for him from the Grenzlandsverwaltungsbehoerde - the Military Frontier Administration Authority - paying 420 gulden a year for him to attend the Joanneum Polytechnic in Graz, Austria. When he has finished, he would then have to serve 8 years in the Military Authority. Tesla left for college with a bag covered with the embroidered designs his mother was famous for. He treasured that bag for the rest of his life.

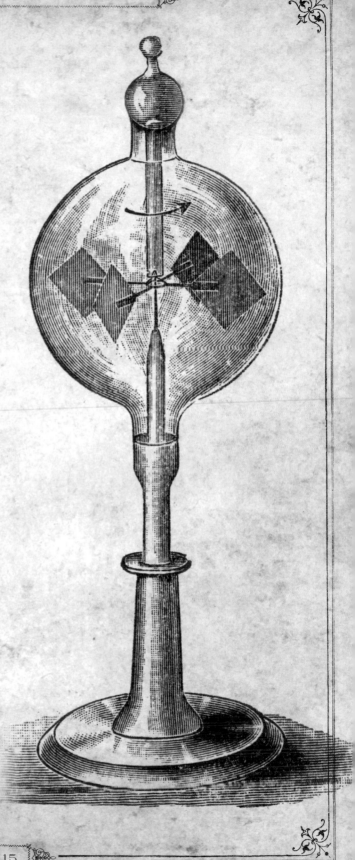

A radiometer, invented by William Crookes (1832 - 1919).

ELECTRICAL SCIENCE IN 1875

The ancient Greeks discovered that you could produce static electricity by rubbing amber with silk. In the 18th century, scientists such as the American Benjamin Franklin (1706 – 90) and the Briton Henry Cavendish (1731 – 1810) made a systematic study of the phenomenon.

In 1791, the Italian Luigi Galvani (1737 – 98) discovered electricity in animal tissue when he saw a frog's leg twitch when touched by two different metals. This led his friend Alessandro Volta (1745 – 1827) to make the first electric battery in 1800.

Danish physicist Hans Christian Ørsted (1777 – 1851) discovered the relationship between electricity and magnetism in 1820 when he saw a compass needle being deflected when an electric current was turned on and off. French physicist André-Marie Ampère (1775 – 1836) developed this into the science of electrodynamics.

In 1831, British scientist Michael Faraday (1791 – 1867) demonstrated the laws of electromagnetic induction, producing a current in a coil by moving a magnet back and forth inside it. This led to the development of both the electric motor and the generator where coils of wire were mounted in a rotor, or armature, within a magnetic field.

As the magnetic effect is only apparent when the current is turned on and off, an electric motor has a commutator – that is, a split ring with electrical contact, or brushes, resting on either side. As the motor turns, contacts switch, reversing the current flow in the coil. Similarly, a generator needs a commutator to prevent the current reversing as the rotor turns.

An early electric motor of the 19th century, driving a mill.

CHAPTER 2

ELECTRIC BRAINWAVES

It has cost me years of thought to arrive at certain results, by many believed to be unattainable, for which there are now numerous claimants, and the number of these is rapidly increasing …

Nikola Tesla[1]

Arriving at the Polytechnic in 1875, Tesla did not study engineering initially. Perhaps in deference to his father, he studied physics and mathematics with the aim of becoming a professor like his Uncle Josif.[2] The Polytechnic had recently bought a Gramme dynamo which physics professor Jacob Pöschl used to teach his students about electric currents. During his lectures, he connected the dynamo to a battery, so it would work as a motor.[3]

While Professor Pöschl was making demonstrations, running the machine as a motor, the brushes gave trouble, sparking badly, and I observed that it might be possible to operate a motor without these appliances. But he declared that it could not be done and did me the honour of delivering a lecture on the subject, at the conclusion he remarked, 'Mr Tesla may accomplish great things, but he certainly will never do this. It would be equivalent to converting a steadily pulling force, like that of gravity into a rotary effort. It is a perpetual-motion scheme, an impossible idea.' But instinct is something which transcends knowledge. We have, undoubtedly, certain finer fibres that enable us to perceive truths when logical deduction, or any other wilful effort of the brain, is futile.[4]

Tesla would go on to make a motor that did without troublesome brushes. It was his first great invention.

EXPERIMENTING WITH THOUGHT

While somewhat intimidated by his professor's authority, Tesla was determined to prove that he was right and 'undertook the task with all the fire and boundless confidence of youth'.[5] To take up the challenge of building a spark-free motor, Tesla switched to the engineering course. However, electrical engineering was in its infancy and the course in Graz concentrated on civil engineering. Consequently, Tesla returned to his thought experiments:

I started by first picturing in my mind a direct-current machine, running it and following the changing flow of the currents in the armature. Then I would imagine an alternator and investigate the progresses taking place in a similar manner. Next I would visualize systems comprising motors and generators and operate them in various ways. The images I saw were to me perfectly real and tangible.[6]

Tesla was a diligent student - for the first year. He worked from 3 am until 11 pm, 7 days a week, taking no holidays. He passed his exams way ahead of his fellow students. But when he went home with his exemplary exam certificates his father was furious. 'That almost killed my ambition,' he wrote.

It was only later, after his father had died, that he discovered letters from his professors telling him to take his son away from the polytechnic, otherwise he would kill himself with overwork.[7]

CAROUSING AND GAMBLING

In his second year at college Tesla gave himself over to carousing and, in his third year, he gave up going to lectures altogether. This led to his scholarship being cancelled. He tried to get another scholarship from the publishers of the pro-Serbian newspaper, *Queen Bee*, calling himself a 'technician' and saying he could speak Italian, French and English, as well as Serbian, Croatian and German.[8] It was refused and he was thrown out of school for gambling and, it was said, 'womanizing'. He disappeared from Graz without a word and friends feared that he had drowned in the river.[9]

In 1878, he re-appeared in Maribor, which was then in the Austrian province of Styria, now in Slovenia. He found work there as a

THE GRAMME DYNAMO

With the rapid development of the telegraph system in the 1840s and 1850s, what was needed was direct current (DC) that flowed in only one direction. This is the type normally produced by batteries.

Even with a commutator, introduced by Parisian instrument-maker Hippolyte Pixii (1808 – 35) in 1832, the current delivered by a generator, while not reversing, was not smooth and constant like that from a battery. It builds to a peak then drops back to zero again. However, Belgian electrical engineer Zénobe-Théophile Gramme (1826 – 1901) demonstrated his Gramme dynamo at the French Academy of Sciences in 1871. By increasing the number of coils on the rotor and the number of sections on the commutator, it could produce a near constant direct current.

Shown at the International Exhibition in Vienna, one Gramme dynamo was connected to another one which acted as a motor. Until then, motors had only been powered by expensive batteries. Gramme's business partner, French engineer Hippolyte Fontaine (1833 – 1910) had demonstrated that power could be transmitted from one place to another without the inefficient shafts, belt, chains or ropes used to connect steam engines to machines – with obvious advantages for industry.

A Gramme dynamo, invented by Zénobe-Théophile Gramme (1826 - 1901).

draftsman in a tool and die shop, though he seems to have spent much of his time playing cards for money.[10] His father, who did not approve of gambling, found out where he was and came to beg him to return to school, this time in Prague.[11]

A few weeks after his father's visit, Tesla was arrested as a vagrant and deported back to Gospic. At his father's church, he met and fell in love with a girl called Anna. Strolling by the river or on long walks back to his hometown of Smiljan, they discussed the future. She wanted a family; he wanted to be an electrical engineer.[12] Then his father fell seriously ill. He died soon after, aged 60, and was given a funeral fitting for a saint.[13]

Tesla continued gambling. One day his mother came to him and gave him a roll of notes, saying: 'The sooner you lose all we possess the better it will be. I know that you will get over it.'[14]

He said: 'I conquered my passion then and there and only regretted that it had not been a hundred times as strong. I not only vanquished but tore it from my heart so as not to leave even a trace of desire. Ever since that time I have been as indifferent to any form of gambling as to picking teeth.'[15] He reported giving up excessive smoking and coffee drinking with similar ease.[16] And he seems to have given up his passion for Anna too.

THE COMING OF THE TELEPHONE

Tesla then honoured his dead father's wishes. Supported by two maternal uncles, he went to Prague University and signed up for courses in mathematics, experimental physics and philosophy. This introduced him to the Scottish philosopher David Hume (1711 - 76) and the idea that human beings were born a blank slate that was shaped through life by sensory perceptions - ideas that would come into play when he

later worked on robotics.[17]

The intellectual ferment of Prague stimulated Tesla and, again, he put his mind to building a new type of electric motor, removing the commutator to eliminate the sparking. Eventually, the money from his uncles dried up. Tesla needed a job and he saw in the newspapers that one of Thomas Edison's agents, Tivadar Puskás, was setting up a telephone exchange in Budapest, having already built one in Paris. Puskás' idea was to build telephone exchanges in major European cities.[18] Until then Alexander Graham Bell had only thought of installing his invention on private lines linking two locations.

However, in Budapest, no work was forthcoming, so Tesla took a government job as a draftsman in the Central Telegraph Office. This bored him and he quit to devote himself full time to inventing. Coming up with no practical idea, he had a nervous breakdown.

A FLASH OF INSPIRATION

Tesla was only rescued from a deep depression by his new friend Anthony Szigeti. One afternoon they were walking in the City Park reciting poetry. 'At that age,' he said, 'I knew books by heart, word for word.'[19] As the sun was setting, he began a passage in German from Goethe's *Faust*. The quote concludes:

Alas the wings that lift the mind no aid
Of wings to lift the body can bequeath me.[20]

Tesla said:

As I uttered these inspiring words the idea came like a flash of lightning and in an instant the truth was revealed. I drew with a stick on the sand the diagram shown six years later in my address before the American Institute of Electrical Engineers, and my companion understood them perfectly. The images I saw

were wonderfully sharp and clear and had the solidity of metal and stone, so much so that I told him, 'See my motor here; watch me reverse it.' I cannot begin to describe my emotions.[21]

The idea Tesla had come up with was using a rotating electric field within the motor.

Although Tesla described his 'Eureka!' moment in his autobiography, he did not patent the alternating current (AC) motor until 1903. He did further experiments on it in 1883 and 1887, and the idea was still not fully worked out when he addressed the AIEE in 1888.

However, Tesla had solved the problem that Professor Pöschl had said was impossible. He was now convinced he was an inventor, and he had made the intellectual breakthrough that would make him rich and famous.

Tesla may have also found inspiration at the works of Ganz and Company in Budapest where AC electrical distribution was being developed.[22] Electricity can be transmitted down wires with less loss at higher voltages. With AC electricity you can step up the voltage - and step it down again - using a transformer. In the Ganz works, engineers found that a metal ball placed on top of a transformer would revolve. Later, Tesla would use this in his Egg of Columbus demonstrations.

PASSPORT TO PARIS

Eventually, Tesla was taken on by Tivadar Puskás' brother Ferenc to work on the new telephone exchange. Tivadar was then in Paris, helping introduce Edison's incandescent lighting system. When the telephone exchange in Budapest was finished, Ferenc sold it to a local businessman and Tivadar offered Tesla and Szigeti jobs in the Edison organization in Paris. Tesla was immediately struck by the 'City of Light'.[23]

I never can forget the deep impression that magic city produced on my mind. For several days after my arrival I roamed through the streets in utter bewilderment at the new spectacle. The attractions were many and irresistible, but, alas, the income was spent as soon as received. When Mr Puskás asked me how I was getting along ... I [replied] 'the last 29 days of the month are the toughest'.[24]

Employed at the Edison works in the suburb of Ivry, Tesla learned a great deal about the practical business of building generators and motors. At the time, little of the basic science had been done and progress was made by trial and error. However, Tesla had the advantage that, unlike the other engineers, he had studied physics and mathematics, and could make calculations.

His schedule, as usual, was unrelenting. He would get up in the morning at 5 am and swim 27 laps of a bathhouse on the Seine. In the evenings he would play billiards with his colleagues.[25] Even then, he would explain his idea for an AC motor, again in the dirt with a stick.

In his spare time, he worked on alternative designs for his flying machine and outlined the specifications for his AC motor in a notebook. It would need three different alternating currents delivered to the motor down six wires at 120° out of phase. This would produce a rotating magnetic field. But he could not get any of Edison's men interested. The business making money at the time was delivering electric light rather than powering motors. The other problem was that, using six wires, rather than the three used in Edison's system, would use much more copper which was a major factor in the cost of new equipment at the time.

Only one Edison man, David Cunningham, saw the potential in Tesla's motor and suggested that they set up a stock company. But Tesla was unfamiliar with the American way of doing business and nothing came of it.[26]

THOMAS ALVA EDISON (1847 – 1931)

Born in Ohio, Edison had little schooling. At the age of 12 he got a job on the railroad where he learned telegraphy. He went on to develop the duplex system that sent two messages at once and a printer that converted electrical signals into letters.

He quit and went into business for himself, developing the quadruplex system, which sent four messages at once for Western Union and their rivals. With the help of his father, he established a laboratory and machine shop at Menlo Park, New Jersey, which became the world's first industrial research facility. There he developed underwater cable for the telegraph, set about improving the primitive telephone developed by Alexander Graham Bell, and inventing the phonograph. This brought him worldwide fame as the *Wizard of Menlo Park*.

He worked on the incandescent light bulb, though battles over patents ensued. He also developed electric motors and generators to power his lighting systems, first on the steamship *Columbia*, then on buildings in New York and London.

A pioneer in motion pictures, he also developed batteries for submarines and the Model T Ford. In all, he took out a world record 1,093 patents and remains the most famous inventor in American history.

Thomas Edison in his Menlo Park laboratory.

ALEXANDER GRAHAM BELL (1847 – 1922)

Edinburgh-born Bell first visited the USA in 1871 where he demonstrated his father's method of teaching speech to the deaf. The following year he opened a school for teachers of the method in Boston.

With young technician Thomas Watson, he set to work on developing ways of using electricity to transmit sound, getting his patent for the telephone in 1876. Hundreds of patent suits followed. Nevertheless, the Bell Telephone Company was established the following year, successfully fighting off suits by two subsidiaries of the Western Union Telegraph Company.

Bell also invented the photophone, which transmitted sound on a beam of light, and the Graphophone, that recorded sound on wax cylinders. He experimented with sonar detection, huge flying kites and hydrofoils, while continuing to find ways to aid the deaf.

Alexander Graham Bell in New York in 1892, making a telephone call to Chicago.

He was still suffering from seeing flashing lights. One night in Paris, he said: 'I felt the positive sensation that my brain had caught fire. I was alight, as though a small sun was located in it and I passed the whole night applying cold compresses to my tortured head.'[27]

However, though painful, these did not worry him. Throughout his life, he said, 'these luminous phenomena still manifest themselves from time to time, as when a new idea opening up possibilities strikes me...'[28]

SECRETS IN STRASBOURG

Tesla was sent as a trouble-shooter to lighting stations in France and Germany. He oversaw the illumination of an opera house in Paris, a theatre in Bavaria and cafés in Berlin. After helping develop an automatic regulator for Edison dynamos, he was sent to fix the illuminations at the central railway station in Strasbourg, which, in the Franco-Prussian War of 1870 - 71, had been taken over by Germany.

There had been a problem there when the wiring had shorted, blowing out a section of the wall during a visit of Kaiser Wilhelm I and a German-speaking engineer was needed to sort it out. In the station's powerhouse, there was a Siemens AC generator. In their spare time, Tesla and Szigeti secretly experimented with the prototype of one of Tesla's AC motors.

'It was the simplest motor I could conceive of,' said Tesla. 'It had only one circuit, and no windings on the armature or the fields. It was of marvellous simplicity.'[29]

The problem was it did not work. The initial trouble was that it used a brass ring that would not magnetize. Steel had to be added in various positions. Then, said Tesla, 'I finally had the satisfaction of seeing rotation effected by alternating current of different phase, and without sliding contacts or commutator, as I had conceived a year before. It was an exquisite pleasure, but not to compare with the delirium of joy following the first revelation.'[30] He had finally proved Professor Pöschl wrong.

PASSAGE TO NEW YORK

Tesla tried unsuccessfully to raise money to back his invention in Strasbourg. He returned to Paris expecting a bonus for his work in Strasbourg which did not materialize. He tried to find financial backing there too, again unsuccessfully. However, he did catch the eye of Charles Batchelor who had been head of the Edison organization in Paris. He was returning to New York to head the Edison Machine Works there and asked Tesla to come with him. To smooth Tesla's passage into the Edison organization Batchelor got a letter of introduction from Tivadar Puskás addressed to Edison, saying: 'I know two great men and you are one of them; the other is this young man.'[31]

Just before leaving for America, Tesla spent time with scientists studying microscopic organisms in drinking water. Having suffered cholera before, he now shunned unpurified water, avoided poor quality restaurants, and scoured the crockery and cutlery before eating. Later he wrote: 'If you would watch only for a few minutes the horrible creatures, hairy and ugly beyond anything you can conceive, tearing each other up with the juices diffusing throughout the water - you would never again drink a drop of unboiled or unsterilized water.'

His uncles again paid for the trip. The journey to New York was not a happy one. His money and most of his belongings were stolen, and there was some sort of mutiny on the ship and Tesla nearly got pushed overboard. On 6 June 1884, Tesla sailed into New York on the *City of Richmond* just as

WHAT IS ALTERNATING CURRENT?

The electricity from a battery, lightning or a Van de Graaf generator that produces a static charge is direct current. It flows in only one direction. Alternating current flows in one direction, then the other. It cycles through peaks and troughs as it changes direction.

With a direct current, when a switch is thrown a magnetic field builds up around the wire, inducing a current in any conductor nearby. This only occurs when the field is building up or when the field collapses when the current is switched off. With an alternating current, the electric current is effectively being switched on and off all the time, inducing an alternating current in the secondary conductor. This property is utilized in an induction motor, where a current is induced in secondary coils on the rotor, and in a transformer, where the voltage is stepped up and down.

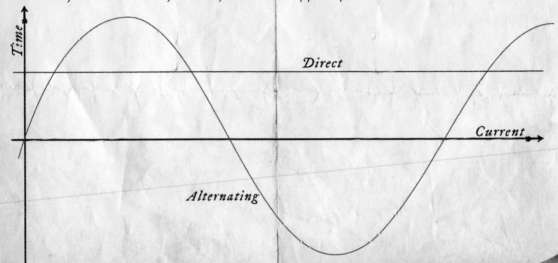

the first stones of the Statue of Liberty's pedestal were being hauled into place. He was 28 years old.

Although he spoke English, Tesla had difficulty making himself understood as the customs officer recorded that he was from Sweden. Tesla then recalled the clerk barking: 'Kiss the Bible. Twenty cents.' His first impressions of America were not good. He wrote:

What I had left was beautiful, artistic, and fascinating in every way. What I saw here was machined, rough, and unattractive. A burly policeman was twirling his stick which looked to me as big as a log. I approached him politely with the request to direct me. 'Six blocks down, then to the left,' he said, with murder in his eyes. 'Is this America?' I asked myself in painful surprise. 'It is a century behind Europe in civilization.' [32]

He went to work immediately. Passing a machine shop, he saw a mechanic who had just given up on trying to fix an electric machine and offered to help. In one version of the story, Tesla did this 'without thought of compensation'; in another, it was one of the machines he had helped design and he charged $20. In either case, legend has that he said: 'I had it running perfectly in an hour.' [33]

Nikola Tesla in 1879, aged 23.

CHAPTER 3

MEETING THOMAS EDISON

I was thrilled to the marrow by meeting Edison who began my American education right then and there. I wanted to have my shoes shined, something I considered below my dignity. Edison said: 'Tesla, you will shine the shoes yourself and like it.' He impressed me tremendously. I shined my shoes and liked it.

Nikola Tesla[1]

Tesla was soon too busy to dwell on the shortcomings of the New World. The dynamos the Edison organization had installed on the SS *Oregon*, then the holder of the Blue Riband for the fastest transatlantic passenger crossing, had broken down, delaying her sailing. Tesla immediately volunteered to make the repairs. He and his crew worked overnight and the *Oregon* set sail the following day for another record-breaking run.

Returning to Edison's Manhattan offices at 5 am, Tesla bumped into Charles Batchelor and Edison, who were just going home. According to Tesla, Edison said: 'Here is our Parisian running around at night.'[2] They had met before in Paris and seem to have got on famously. Edison recalled:

> Oh, he's a great talker, and, say, he's a great eater too. I remember the first time I saw him. We were doing some experimenting in a little place outside Paris, and one day a long, lanky lad came in and said he wanted a job. We put him to work thinking he would soon tire of his new occupation for we were putting in twenty to twenty-four hours a day then, but he stuck right to it and after things eased up one of my men said to him: 'Well, Tesla, you've worked pretty hard, now I'm going to take you into Paris and give you a splendid supper.' So he took him to the most expensive café in Paris – a place where they broil an extra thick steak between two thin steaks. Tesla stowed away one of those big fellows without any trouble and my man said to him: 'Anything else, my boy? I'm standing you a treat.' 'Well, if you don't mind, sir,' said my apprentice, 'I'll try another steak.' After he left me he went into other lines and has accomplished quite a little.[3]

At their first meeting in New York, Tesla explained that he had just returned from the *Oregon* and had repaired both the dynamos. Edison said nothing and walked away. When he was a little distance away, Tesla saw him turn to Batchelor and say: 'Batchelor, this is a damn good man.'[4]

EMULATING EDISON

Plainly Tesla had impressed Edison and began work at the Edison Machine Works on 8 June 1884, only two days after arriving in America. He worked hard. 'For nearly a year, my regular hours were from 10.30 am to 5 o'clock the next morning without a day's exception,'[5] he said.

Possibly thinking that Tesla came from Transylvania, Edison asked whether Tesla had ever tasted human flesh. Tesla was also appalled at Edison's 'utter disregard of the most elementary rules of hygiene',[6] but enquired what the great man's diet consisted of. 'You mean to make me so all-fired smart?' said Edison. Tesla nodded.

Edison replied, perhaps in jest, that he ate a daily regimen of Welsh Rarebit as 'it's the only breakfast guaranteed to renew one's mental faculties after the long vigils of toil'. Tesla began to do the same despite a protesting stomach. But his admiration was undiminished.[7] Tesla wrote in his autobiography:

> I was amazed at this wonderful man who, without early advantages and scientific training, had accomplished so much. I had studied a dozen languages, delved in literature and art, spent my best years in libraries reading all sorts of stuff that fell into my hands, from Newton's *Principia* to the novels of Paul de Kock, and felt that most of my life had been squandered.[8]

The admiration was mutual. According to Tesla, Edison told him: 'I have had many hard-working assistants, but you take the cake.'[9]

However, Tesla still took the time to enjoy a good meal - the *table d'hôte* at a restaurant in Greenwich Village with a bottle of red wine

PEARL STREET POWER STATION

The first electric lights were arc-lamps that gave off light from electric sparks. But in 1879, Edison came up with the improved incandescent lamp. Arc lamps had been connected in series – if one failed, all of them went out. Edison connected his lamps in parallel, so each faulty bulb could be replaced individually. This had created an astonishing demand for electric power. Edison built his first DC power station on Pearl Street in lower Manhattan in 1882, initially to power 400 lamps owned by 85 customers. It quickly became a monopoly and by 1884 it was serving 508 customers with 10,164 lamps. The electricity was carried above ground on poles with dozens of crooked crossbeams supporting sagging wires. The exposed electrical wiring was a constant danger and unsuspecting children climbing the poles would suffer lethal electric shocks. In spite of the perils, wealthy New Yorkers rushed to have their homes wired, the most important being the banker, J.P. Morgan, who had invested heavily in Edison.

A gentleman explains the first Edison Electric Lighting Station at Pearl Street in lower Manhattan in the 1880s.

- and play billiards, a game he had mastered as a student. According to Edison's personal secretary Alfred O. Tate: 'He played a beautiful game. He was not a high scorer but his cushion shots displayed a skill equal to that of a professional exponent of this art.'[10]

AVOIDING THE AC SUBJECT

Tesla set about redesigning Edison's dynamos, replacing their long magnets with more efficient short cores, claiming that they gave three times the output for the same amount of iron. He kept quiet about his AC motor though, perhaps recalling the indifference of Edison's men in Paris. Once, though, he was tempted to bring up the subject with Edison himself.

'It was on Coney Island,' he said, 'and just about as I was going to explain it to him, someone came and shook hands with Edison. That evening, when I came home, I had a fever and my resolve rose up again not to speak freely about it to other people.'[11] Otherwise Tesla and Edison got on well enough. Tesla told the tale of visiting Edison's office at 65 Fifth Avenue, when the great man was playing a game guessing weights.

'Edison felt me all over and said: "Tesla weighs 152 pounds to the ounce" - and he guessed it exactly,'[12] Tesla recalled. He asked how Edison could guess his weight so accurately and was told: 'He was employed for a long time in a Chicago slaughter house where he weighed thousands of hogs every day.'[13]

Tesla would occasionally dine with Edison, Batchelor and others of the company's top brass in a restaurant across the road from the showroom at 65 Fifth Avenue where they would swap stories and tell jokes. Afterwards they would go to a billiard room where Tesla would impress them with his bank shots and his vision of the future.

ARC LIGHTING

Companies that had grown up making arc lights were now moving into incandescent lighting, robbing Edison of valuable contracts. He struck back by going into arc lighting which was more suitable for street lighting or illuminating large spaces. He filed an arc-lamp patent in June 1884 and left Tesla to work out the details. Tesla completed the job, but his system was shelved when Edison made a deal with a dedicated arc-lighting company and, by then, larger incandescent bulbs suitable for lighting larger spaces had been developed. Tesla felt cheated.

'The manager had promised me $50,000,' he wrote, 'but when I demanded payment, he merely laughed, saying "You are still a Parisian! When you become a fully-fledged American, you will appreciate an American joke."'[14]

As it was, Tesla could not even get his salary of $18 a week increased to a modest $25. This was a painful shock and he resigned. Later in life, Tesla re-assessed his opinion of Edison. When the Wizard of Menlo Park died in 1931, Tesla said:

If he had a needle to find in a haystack he would not stop to reason where it was most likely to be, but would proceed at once with the feverish diligence of a bee, to examine straw after straw until he found the object of his search ... I was almost a sorry witness of his doings, knowing just a little theory and calculation would have saved him ninety per cent of the labour ... Trusting himself entirely to his inventor's instinct and practical American sense ... the truly prodigious amount of his actual accomplishments is little short of a miracle.[15]

INDEBTED TO EDISON

Despite the rift between the two men, Tesla was indebted to Edison. With Edison's former patent attorney Lemuel Serrell, Tesla began patenting improvements to arc lights and dynamos.

In Serrell's office, Tesla met B.A. Vail and Robert Lane. They set up the Tesla Electric Light and Manufacturing Company. Tesla tried to interest them in his AC motor, but they were only interested in arc lighting. Together they set about lighting the streets and factories in Rahway, New Jersey, Vail's hometown. Meanwhile Tesla used the patents he had been granted to buy shares.[16]

When the electrification of Rahway was completed, *Electrical Review* featured it on the front page of its 14 August 1886 issue. It was so successful that Vail and Lane decided to run the utility, leaving no role for Tesla. He was bounced from the company leaving him with nothing but 'a beautifully engraved certificate of stock of hypothetical value'.[17] He could not even use his own inventions as patents had been assigned to the company. It was the 'hardest blow I ever received,'[18] he said. In the winter of 1886 - 87, he was forced to dig ditches.

I lived through a year of terrible heartaches and bitter tears, my suffering being intensified by my material want. My high education in various branches of science, mechanics and literature seemed to me like a mockery.[19]

RESCUED BY A PATENT

In 1884, Edison had been experimenting with ways of producing electricity from burning coal or gas. It ended when an explosion blew the laboratory's windows out. However, Tesla figured out a simpler - and safer - way to do it, and in March 1886, he applied for a patent for his thermo-magnetic motor.

TESLA'S THERMO-MAGNETIC MOTOR

When iron magnets are heated they lose their magnetic strength. Tesla devised a small motor with one fixed magnet and a second moving magnet attached to a flywheel and a pivot arm that pushed against a spring. At room temperature, the attraction between the two magnets was enough to compress the spring and pull the moving magnet into the flame of a Bunsen burner. This would heat up the magnet. It would lose strength. The spring would then push the magnet back, turning the flywheel. With the magnet now out of the flame, it cooled down. Its magnetic strength was restored and the cycle began all over again.

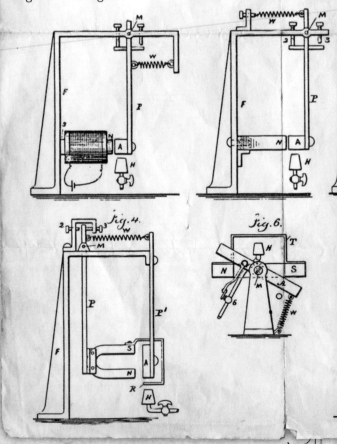

Patent application for Tesla's thermo-magnetic motor, March 1886.

While Tesla was digging ditches, he told his foreman about his inventions. It seems that the foreman had been digging the ditches for the underground cables that connected Western Union's head office with the stock and commodity exchanges, and knew Alfred S. Brown (1836 - 1906) who was superintendent of Western Union's New York Metropolitan District.[20]

Brown probably knew of Tesla from the article in *Electrical Review* and was impressed by his thermoelectric motor and his AC inventions.

TESLA'S PYROMAGNETIC GENERATOR

Tesla took a large horseshoe magnet. Across its poles ran a number of hollow iron tubes which were magnetized and had coils of wire wrapped around them. Under the centre of the tubes was a firebox. Above there was a boiler. The coal fire in the firebox heated the iron tubes until they reached about 600°C (1,112°F) and glowed a dull red. As the tubes heated up, they would lose their magnetism and the collapsing magnetic field would induce an electric current in the coils. Then a valve was opened and steam, at 100°C (212°F), would circulate through the tubes, cooling them. As the magnetism in the tubes was restored, electric current would again be induced in the coils.[21]

N. TESLA.
PYROMAGNETO ELECTRIC GENERATOR.

No. 428,057.   Patented May 13, 1890.

Inventor
Nikola Tesla

Patent application for Tesla's pyromagnetic generator.

THE EGG OF COLUMBUS

Eager to exploit Tesla's ideas, Brown contacted Charles F. Peck, a wealthy lawyer from New Jersey. However, Peck knew of the general prejudice against AC and refused even to witness some tests.

'I was discouraged,' Tesla said. But then he remembered the 'Egg of Columbus'. The story goes that Columbus was having dinner with some Spanish nobles who mocked him. So he challenged them to stand an egg on its end. They all tried and failed. Then he took the egg, tapped it lightly on one end, cracking the shell and denting it, so it would stand upright. As a result he was granted an audience with Queen Isabella and won her support for his voyage.

Tesla told Peck that he could go one better. He would make an egg stand on its end without cracking the shell. If he could trounce Columbus, Tesla asked, could he count on Peck's support? Peck said he had no crown jewels to pawn, but he would help.[22]

After the meeting, Tesla took a hard-boiled egg to a blacksmith and got him to cast one in copper. Then he placed four coils under the top of a wooden table to create a magnetic rotating field. When he turned on the current, the egg began to spin. Growing faster, it ceased to wobble and stood on its end. Peck was impressed. Not only had Tesla gone one better than Columbus, he had demonstrated the principle of his AC motor.

TESLA ELECTRIC COMPANY

Together Tesla, Brown and Peck formed the Tesla Electric Company in April 1887. Tesla would get a third share of any money generated. Brown and Peck would split another third, and a third would be reinvested. Tesla also received a salary of $250 a month, while Brown and Peck would cover the cost of the patents. The following month, Szigeti came to New York to work as Tesla's assistant.

They set up a laboratory at 89 Liberty Street above a printing company. During the day the printing shop used a steam engine to power its presses. At night it provided power for Tesla's experiments.[23]

First Tesla developed his thermomagnetic motor into the pyromagnetic generator, using the same principles. Tesla believed this was a great invention, but it did not work very well and his patent application was denied. Nevertheless, Peck encouraged him to continue inventing and his mind turned back to the AC motor he built in Strasbourg.

This time, instead of a single coil, he used four coils wound around a laminated ring. Two separate AC currents were fed to the pairs of coils on opposite sides. If these two currents were 90° out of phase - that is, when one was at its maximum positive value, the other was at its maximum negative value - a rotating electrical field was produced. To Tesla's delight, the rotor - initially a shoe polish tin balanced on a pin - began to spin.[24]

From this prototype, Tesla and Szigeti produced two full-scale motors. They were, Tesla said, 'exactly as I had imagined them. I made no attempt to improve the design, but merely reproduce the pictures as they appeared to my vision and the operation was as I expected.'[25]

Patents were applied for and issued on 1 May 1888. The motors were then tested for their efficiency by Professor William Anthony (1835 - 1908) of Cornell University and, on 15 May, Tesla delivered his ground-breaking paper *A New Alternating Current Motor* to the American Institute of Electrical Engineers (AIEE).

Nikola Tesla in 1885, aged 29.

CHAPTER 4

THE WESTINGHOUSE CORPORATION

There were many days when [I] did not know where my next meal was coming from. But I was never afraid to work, I went where some men were digging a ditch … [and] said I wanted to work. The boss looked at my good clothes and white hands and laughed to the others … but he said, 'All right. Spit on your hands. Get in the ditch.' And I worked harder than anybody. At the end of the day I had $2.

Nikola Tesla[1]

Power stations had sprung up across America and Europe to provide electric light at night. In the 1880s the owners saw electric motors as a way to sell power to factories and streetcar lines during the day. However, most of the power stations produced DC and Brown and Peck were a little dubious of Tesla's fixation with AC.

Other inventors had used AC to power arc lights. This was particularly popular in Europe where experimenters found they could raise or lower the voltage of an alternating current using primitive transformers. Engineers at the Ganz Company found that, at a high voltage, electricity could be distributed over long distances using thin copper wires. Then, to make it safe to use in the home, it would be stepped down using a transformer.

WESTINGHOUSE AND AC

In 1884, George Westinghouse became interested in electric lighting and hired the inventor William Stanley Jr (1858 - 1916), who had invented an incandescent lamp and a self-regulating dynamo. At first, Westinghouse thought of developing a DC system, but abandoned it as the market was already overcrowded.

For a DC system to be profitable, numerous small power stations had to be situated near to the homes and factories they served. Westinghouse saw that an AC system could provide power to towns and cities where the population was spread out, reaching a market that DC systems could not serve. There would also be considerable economies of scale by using an AC system where the voltage could be stepped up and distributed over a wider area, serving more customers. The disadvantage was that no one had yet developed a meter to measure how much each individual consumer used, nor

had a serviceable AC motor been developed to provide power for factories and streetcars. Besides, Edison had dismissed AC as 'not worth the attention of practical men'.[2]

In 1885, Westinghouse imported a transformer made by Lucien Gaulard and John Gibbs in London which Stanley began experimenting with. By the time Tesla presented his paper to the AIEE in May 1888, Westinghouse had already sold more AC power stations than all the other companies providing DC power put together. In response, Edison began warning that AC power was dangerous.

TESLA DEMONSTRATES AC

In his paper to the AIEE, Tesla sought to demonstrate the superiority of AC once and for all. He told the audience gathered in the lecture hall at Columbia University:

The subject which I now have the pleasure of bringing to your notice is a novel system of electric distribution and transmission of power by means of alternate currents, affording peculiar advantages, particularly in the way of motors, which I am confident will at once establish the superior adaptability of these currents to the transmission of power and will show that many results heretofore unattainable can be reached by their use; results which are very much desired in the practical operation of such systems and which cannot be accomplished by means of continuous currents ...

In our dynamo machines, it is well known, we generate alternate currents which we direct by means of a commutator, a complicated device and, it may be justly said, the source of most of the troubles experienced in the operation of the machines. Now, the currents so directed cannot be utilized in the motor, but they must – again by means of a similar unreliable device – be reconverted into their original state of alternate currents. The function of the commutator is

THE TRANSFORMER

The transformer uses the same principles of electromagnetic induction employed in electric motors and generators. Two coils of wire are wound around a single iron core. When an electrical current is passed through one of them, it magnetizes the iron core. This, in turn, induces an electric current in the other one. The voltage is stepped up or stepped down according to the ratio of the number of turns of wire in each coils. However, induction only works when the electrical current is being switched on and off again, so an alternating current rather than a direct current must be used.

The transformer is a vital component of any power distribution system as transmission losses are much smaller when the voltage is higher – as less current is needed to convey the same amount of energy. So electricity generated at a power station is stepped up in voltage using a transformer before it reaches the transmission lines. Then, at its destination, it is stepped down for use in the home or factory.

Westinghouse AC generator. The world's first single-phase AC power transmission system. Built in 1895, by Nikola Tesla and George Westinghouse at Ames Hydroelectric Plant, Telluride, Colorado (photo c. 1900).

entirely external, and in no way does it affect the internal working of the machines. In reality, therefore, all machines are alternate current machines, the currents appearing as continuous only in the external circuit during their transit from generator to motor.[3]

After the lecture, Tesla demonstrated that his AC motor could be instantly reversed. He provided precise calculations on how the speed and power of the motor could be determined, and he showed how his system could be married up to DC apparatus.

Professor Anthony said that, in his test, he had found Tesla's motors 50 - 60 per cent more efficient than DC models. Then arc-lighting pioneer Elihu Thomson (1853 - 1937) stepped forward to say that he had already developed an AC motor. However, his still used a commutator and was consequently inefficient. Tesla pointed this out and earned himself a life-long enemy, while he himself was catapulted to fame.

SELLING THE AC MOTOR PATENTS

Brown and Peck now invited bids on Tesla's patents. Westinghouse expressed an interest, only to be told that a Californian syndicate had offered $200,000 plus $2.50 per horsepower for each motor installed. The terms were monstrous, but without the patents it would be impossible for Westinghouse to develop a motor of his own.

'The price seems rather high, but if it is the only method for operating a motor by alternating current, and if it is applicable to a streetcar work,' Westinghouse wrote, 'we can unquestionably easily get from the users of the apparatus whatever tax is put upon it by the inventors.'[4]

On 7 July 1888, Westinghouse agreed to pay $25,000 in cash, $50,000 in notes and a royalty of $2.50 per horsepower for each motor. Westinghouse also guaranteed that the royalties would be at least $5,000 in

the first year, $10,000 in the second year and $15,000 a year from then on. Brown and Peck were also reimbursed the money they had paid out on the development of the motors. Over 10 years Tesla, Brown and Peck stood to make $200,000 and $315,000 over the 17-year lifetime of the patents.[5]

As Brown and Peck had negotiated this shrewd deal, as well as bearing the financial risk of developing the motors, Tesla gave them five-ninths of the deal, keeping four-ninths for himself.[6]

MOVING TO PITTSBURGH

Later that month, Tesla took the train to Pittsburgh to put his motors into production at Westinghouse's factory, where he met the great man himself. Later Tesla paid tribute to George Westinghouse, saying on his death:

The first impressions are those to which we cling most in later life. I like to think of George Westinghouse as he appeared to me in 1888, when I saw him for the first time. The tremendous potential energy of the man had only in part taken kinetic form, but even to a superficial observer the latent force was manifest. A powerful frame, well proportioned, with every joint in working order, an eye as clear as a crystal, a quick and springy step – he presented a rare example of health and strength. Like a lion in a forest, he breathed deep and with delight the smoky air of his factories. Though past 40 then, he still had the enthusiasm of youth. Always smiling, affable and polite, he stood in marked contrast to the rough and ready men I met. Not one word which would have been objectionable, not a gesture which might have offended – one could imagine him as moving in the atmosphere of a court, so perfect was his bearing in manner and speech. And yet no fiercer adversary than Westinghouse could have been found when he was aroused. An athlete in ordinary life, he was transformed into a giant when confronted

GEORGE WESTINGHOUSE (1846 – 1914)

George Westinghouse was a descendent of the aristocratic Russian von Wistinghousen family. His father was also an inventor with six patents for farming machinery to his name. In his father's machine shop in Schenectady, New York, the young Westinghouse experimented with batteries and Leyden jars – glass jars coated with metal foil, used for storing electrical charge. At 15, he made his first invention, a rotary steam engine. After serving in both the US Army and Navy during the American Civil War, he attended the nearby Union College, but soon dropped out. In 1865, he patented his rotary engine, and a device for putting derailed freight cars back on the tracks. Soon after, he designed a reversible cast-steel frog which prolonged the life of railroad track switches.

Having been involved in a near collision on the railway, he put his mind to improving the braking system which, until then, had depended on a brakeman on every carriage. His first system, using steam, proved impractical. But then in 1869 he came up with air-brakes that soon became standard in the US and Europe.

He then worked on signalling, devising an electrical system. With the aid of Tesla, Westinghouse entered the 'War of the Currents', championing AC against Edison's DC system. By 1889, the Westinghouse Electric Corporation was a global company, employing over 500,000 people. However, in the financial panic of 1907, he lost control of the companies and retired altogether in 1911.

with difficulties which seemed insurmountable. He enjoyed the struggle and never lost confidence. When others would give up in despair he triumphed. Had he been transferred to another planet with everything against him he would have worked out his salvation. His equipment was such as to make him easily a position of captain among captains, leader among leaders. His was a wonderful career filled with remarkable achievements. He gave to the world a number of valuable inventions and improvements, created new industries, advanced the mechanical and electrical arts and improved in many ways the conditions of modern life. He was a great pioneer and builder whose work was of far-reaching effect on his time and whose name will live long in the memory of men.[7]

TESLA'S TASK

Founded in 1886, the Westinghouse Electric Company made $800,000 in 1887 and over $3 million in 1888, despite expensive legal battles with Edison. Even so, Tesla took no salary while he worked there. He did this on principle, he said. Since he devoted himself to scientific research, he would never accept fees or compensations for his professional services. However, after a year, he was given a 150 shares of capital stock and he was given $10,000 when he discovered that Bessemer steel made a better transformer core than soft iron.

In New York, he had lived in a garden apartment. In Pittsburgh, he got his first taste of living in grand hotels.

DYING IN THE NAME OF SCIENCE

Under the headline: 'Died for Science's Sake – A Dog Killed With The Electric Current', *The New York Times* of 31 July 1888 reported on one of Harold P. Brown's demonstrations. In Professor Chandler's lecture room at Columbia's School of Mines, Brown told an invited audience that he represented no company and no financial or commercial interest. He also maintained that he had proved by repeated experiments that a living creature could stand shocks from a continuous current much better than from an alternating current. He had applied 1,410 volts of DC to a dog without fatal results, but had repeatedly killed dogs with 500 to 800 volts of alternating current.

Brown then brought out a Newfoundland mix weighing 76 pounds (34 kg). The dog was muzzled and tied down inside a wire cage. The newspaper reported:

Mr Brown announced that he would first try the continuous current at a force of 300 volts [DC]. When the shock came the dog yelped and then subsided. A relay has been attached to the apparatus, which shut off the current almost as soon as it was applied. When the dog got 400 volts he struggled considerably, still yelping. At 700 volts he broke his muzzle and nearly freed himself. He was tied again. At 1,100 volts his body contorted with the pain of the brutal experiment. His resistance to the current then dropped from 15,000 to about 2,500 ohms.

'He will be less trouble,' said Mr Brown, 'when we try the alternating current. As these gentlemen say, we shall make him feel better.' It was proposed the dog be put out of his misery at once. This was done with an alternating current of 330 volts killing the beast.[8]

When Brown brought out another dog, an agent from the American Society for the Prevention of Cruelty to Animals stepped in. Meanwhile, the assembled electricians said that the dog had been weakened by the DC current before the AC was applied. But Brown insisted that the only places AC should be used were 'the dog pound, the slaughter house and the State prison'.[9]

Tesla's task in Pittsburgh was to adapt his motors, which ran best at 50 or 60 cycles per second, to the 133 cycles per second used by the Gaulard-Gibbs transformer at the 120 power stations Westinghouse had already set up. They used this higher frequency to prevent the lights flickering. Since Westinghouse's chief electrician, Oliver B. Shallenberger (1860 - 98), had adapted an electrical meter to run at this frequency, it was reasonable to believe that Tesla's motors could be adapted too. Tesla also had the problem of adapting his motors to run on the two-wire system used by Westinghouse power stations, rather than the four or six wires used to provide out-of-phase AC current to his prototypes.

ENCOUNTERING NIKOLA TESLA

At Westinghouse, Tesla's assistant, was to be Charles Scott (1864 - 1944) who admitted that he had only learned that there was such a thing as alternating current in 1887 after reading an article written by William Stanley in *Electrical World*.

'I had graduated from college two years earlier, and I wondered why I had not heard of such things from my professors,'[10] said Scott. Now, a year later, he was to meet Tesla himself. 'There he came, marching down the aisle with head and shoulders erect and with a twinkle in his eye. It was a great moment for me,'[11] he said. Scott would later become a Professor of Engineering at Yale, but in Pittsburgh he was 'Tesla's wireman ... preparing and making tests'.[12] He recalled:

It was a splendid opportunity for a beginner, this coming in contact with a man of such eminence, rich in ideas, kindly and friendly in disposition. Tesla's fertile imagination often constructed air castles which seemed prodigious. But, I doubt whether even his extravagant expectations for the toy motor of those days measured up to actual

realization ... for the polyphase system which it inaugurated ... exceeded the wildest dreams of the early days.[13]

THE WAR OF THE CURRENTS

With the successful introduction of Westinghouse's AC system, Edison, who was still wedded to DC, was on the back foot and a full-scale industrial war broke out. He launched an all-out propaganda onslaught on the dangers of alternating current. Westinghouse, who would be the eventual victor, said:

I remember Tom [Edison] telling them that direct current was like a river flowing peacefully to the sea, while alternating current was like a torrent rushing violently over a precipice. Imagine that! Why they even had a professor named Harold Brown who went around talking to audiences ... and electrocuting dogs and old horses right on stage, to show how dangerous alternating current was.[14]

REPLACING THE HANGMAN

Having collected a list of over 80 casualties, Harold P. Brown was concerned over the safety of electricity. At the School of Mines, part of Columbia University in New York City, he began experimenting with animals to show that AC was more dangerous than DC. In December 1888, Edison brought Brown out to Menlo Park to electrocute animals with AC as part of his propaganda war.

A number of cities began using electricity to clear their dog pounds. New York went a step further and set up a commission, under the auspices of the Medico-Legal Society of New York to see whether electricity could be used for capital punishment. Brown became the official state electrical execution expert. The guinea pig was to be William Kemmler, a 30-year-old alcoholic who had killed his common-law wife with a hatchet. Edison testified to the committee. The following day the *New-York Daily Tribune* carried the headline: 'Edison Says It Will Kill, The Wizard Testifies As An Expert In The Kemmler Case, He Thinks An Artificial Current Can Be Generated Which Will Produce Death Instantly And Painlessly In Every Case - One Thousand Volts Of An Alternating Current Would Be Sufficient.'[15] The *Electrical Review* detailed Edison's proposed method:

> *He proposes to manacle the wrists, with chain connections, place ... the culprit's hand in a jar of water diluted with caustic potash and connect therein ... to a thousand volts of alternating current ... place the black cap on the condemned, and at a proper time close the circuit. The shock passes through both arms, the heart and the base of the brain, and death is instantaneous and painless.*[16]

Brown surreptitiously bought some Westinghouse motors and began experimenting with larger animals. In Edison's laboratory he 'Westinghoused' 24 dogs, bought from local children at 25 cents each, while Edison himself 'Westinghoused' two calves and a horse. After this demonstration, *The New York Times* reported: 'The experiments proved the alternating current to be the most deadly force known to science, and that less than half the pressure used in this city for electric lighting by this system is sufficient to cause instant death. After Jan. 1 the alternating current will undoubtedly drive the hangman out of business in this State.'[17]

ELECTRICAL CHALLENGE

George Westinghouse wrote to *The New York Times* in protest. Brown responded by using the letters page to draw attention to Westinghouse's 'pecuniary interests' in 'death-dealing alternating current' and issued a challenge:

> *I therefore challenge Mr Westinghouse to meet me in the presence of competent electrical experts and take through his body the alternating current while I take through mine a continuous current. The alternating current must have not less than 300 alternations per second (as recommended by the Medico-Legal Society). We will commence with 100 volts, and will gradually increase the pressure by 50 volts at a time, with each increase, until either one or the other has cried enough, and publicly admits his error.*[18]

Westinghouse did not reply, though Brown was later denounced by the *New York Sun*. Under the headline 'For Shame, Brown', the newspaper revealed that he had been 'paid by one electric company to injure another'.[19] Brown protested: 'I am exposing the Westinghouse system as any right-minded man would expose a bunco starter or the grocer who sells poison where he pretends he sells sugar.'[20] But his protests did no good.

FACING THE ELECTRIC CHAIR

William Kemmler's execution went ahead on 6 August 1890. It was neither instantaneous nor painless. 'To the horror of all present, the chest began to heave, foam issued from the mouth, and the man gave every evidence of reviving,'[21] said *Electrical Review*.

A doctor present told *The New York Times* he would rather have seen ten hangings. Westinghouse said they would have done better with an axe, while Edison blamed the doctors. They had applied the current to the top of the head, though hair was not conductive, and they had not put his

INSIDE AN INDUCTION MOTOR

An electrical motor works through the interaction of two sets of magnets – one stationary, the stator, and one able to move freely, the rotor. For practical motors, electromagnets are used as they don't weaken over time. Permanent magnets do. So both the stator and rotor are essentially coils of wire around metal cores. However, even electromagnets have no reason to set the rotor spinning. But magnetic attraction will draw a north pole on the rotor towards a south pole on the stator. To keep the motor turning, the polarity of the electromagnets has to keep on switching. In early DC motors this was done by a split ring commutator supplying current to the rotor. Tesla's genius was that he realized that you did not have to send electricity to the electromagnets on the rotor at all. If you supplied alternating current to the coils on the stator, the magnetic field created would induce an electric current in the coils of the rotor. The magnetic field produced would oppose that on the stator and the motor would turn.

Essentially, a dynamo or generator is a motor worked in reverse. With a motor you supply electricity and get a mechanical turning force. In a dynamo you supply the mechanical turning force and get electricity. Add in the transformer and Tesla had created a complete system of generating, transmitting and utilizing power.

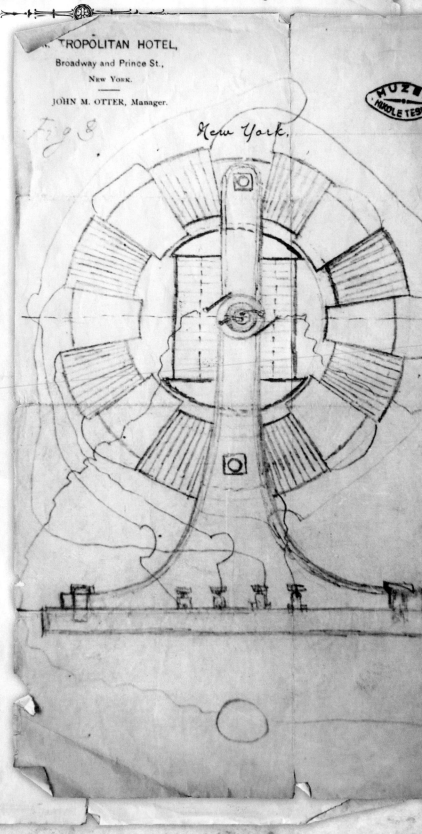

A diagram of the inner workings of an induction motor, drawn by Nikola Tesla on Metropolitan Hotel notepaper.

hand in a jar of water. However, next time they would get it right, he said. Brown was notable by his absence.[22]

But the damage had been done. Backers began to pull their money out of Westinghouse and work on Tesla's induction motor was abandoned. But Tesla still had faith in his invention and agreed to remove the royalty clause from the contract if work resumed.

'George Westinghouse was, in my opinion, the only man on the globe who could take my alternating current system under the circumstances then existing and win the battle against prejudice and money power,' Tesla said. 'He was a pioneer of imposing stature and one of the world's noblemen.'[23]

After 2 years, work on Tesla's motors resumed. The young engineer Benjamin Lamme examined the patents and the prototypes, and concluded that Tesla had exhausted all the possibilities of adapting them to run at higher frequencies. He managed to talk his superiors round. Westinghouse had to go over to 60 cycles per second - the frequency of alternating current used to this day in the US - and the company simply announced that a young engineer named Lamme had discovered the efficiencies of lower frequencies.[24]

Cross of the Légion d'Honneur for his achievements.

Meanwhile Tesla met Norwegian scientist Vilhelm Bjerknes (1862 - 1951) whose study of electrical resonance was vital in the development of radio. While dining out in Paris with French engineer and physicist André Blondel, he also ran across the famous French actress Sarah Bernhardt (1844 - 1923) when she dropped her lace handkerchief near his dining table and Tesla rushed to hand it back to her. Legend has it that their eyes met with a burning intensity. The *Electrical Review* remarked that he may be 'invulnerable to Cupid's shafts',[25] but Sarah Bernhardt may have been an exception. Tesla said much later, that it was a scarf he had picked up, not a handkerchief and he did not return it. He kept it, without washing it, for the rest of his life.

While Edison continued his tour of Europe, being feted everywhere he went, Tesla paid a short visit to his family, then returned to New York, where he opened a laboratory on Grand Street. There he began work on lighting and radio transmission. 'I was not free in Pittsburgh,' he said. 'I was dependent and could not work ... When I [left] that situation, ideas and inventions rushed through my brain like Niagara.'[26]

TESLA TRAVELS THE WORLD

In 1889, Tesla left Pittsburgh and went to Paris for the Universal Exposition of 1889 where the Eiffel Tower made its first appearance. Edison was there too. He had been given a one-acre site to display his inventions and the latest - the phonograph - was causing a sensation. With his new wife, 24-year-old Mina, Edison had lunch with Alexander Eiffel in his apartment at the top of the tower. He also visited Louis Pasteur in his laboratory and received the Grand

MAKING OUTLANDISH CLAIMS

When Edison was in London, visiting his power stations, an engineer named Sebastian Ziani de Ferranti (1864 - 1930) was building an AC power station in Deptford, south London, that was able to transmit electricity at 11,000 volts to central London, seven miles away. Meanwhile, Tesla claimed that he had a system that could 'place a 100,000 horsepower on a wire and send it 450 miles in one direction to New York City, the metropolis of the East, and 500 miles in the

other direction to Chicago, the metropolis of the West'.[27]

This outlandish claim was greeted like those of the current conmen John Ernst Worrell Keely (1837 - 98), who was jailed in 1888 for contempt of court after having claimed to have invented a perpetual-motion machine, and Walter Honenau, who tried to sell pills that, he said, turned water into petrol. However, Tesla's claim turned out, just a few years later, to be true.

Tesla also continued helping Westinghouse with their development of his motor, again taking no payment apart from the equipment Westinghouse provided to furnish his new lab.

Tesla stayed in touch with the men in his family by letter, but usually only sent cheques to his sisters. He also made an effort to pay back all the money his uncles had spent on him. Uncle Petar had risen to become a Metropolitan - the Orthodox equivalent of a Cardinal - while Uncle Pajo responded by sending fine wines as Tesla often complained of the poor quality of wine in the US. He had little time to enjoy them though. He was working seven days a week, stopping only to freshen up in the hotel that had now become his home, or to keep some pressing appointment.

EXPERIMENTAL PHYSICS

Rather than working as an engineer, Tesla was now more of an experimental physicist. His close friend, electrical engineer Thomas Commerford Martin (1856 - 1924), president of the AIEE 1887 - 88 and editor of *Electrical World*, was writing an article on Tesla. During their interviews, Tesla would intersperse his diatribes on how his Serbian family had fought off the diabolical Turks with new theories on electromagnetism and the structure of light.

After Martin's article was published in February 1890, a meeting of the AIEE was devoted to Tesla's AC system. Experiments in the long-distance transmission of AC current were then being conducted in Germany and Switzerland.[28] Westinghouse was opening a hydroelectric plant at a mining camp in Telluride, Colorado, using Tesla's AC system and the International Niagara Commission announced that it was looking at the best way to exploit the Niagara Falls.

WHIRLING THROUGH ENDLESS SPACE

Tesla agreed to present his work on high-frequency phenomena to the AIEE. The meeting would be open to the public and the journalist covering the symposium for *Electrical World* also managed to sell a piece to the prestigious mainstream magazine *Harper's Weekly*. It said that Tesla had gone beyond noted European physicists such as Professor Heinrich Hertz in his understanding of the electromagnetic theory of light. [29]

He also put on a number of spectacular demonstrations. In one, light streamed from a wire attached at one end to a coil. In another, a fine thread of platinum wire inside a glass bulb span, forming a funnel of light. He produced light bulbs that worked with just one wire attached, or with none at all, and generated huge sparks and electric flames. Electricity, he showed, would run to earth.

As a finale, Tesla ran tens of thousands of volts of AC through his body to light up light bulbs and shoot sparks from his fingertips to show that alternating current was not a killer when handled properly. *Electrical World* said: 'Exhausted tubes ... held in the hand of Mr Tesla ... appeared like a luminous sword in the hand of an archangel representing justice.'[30] Tesla concluded his lecture, saying:

We are whirling through endless space with an inconceivable speed. All around us everything is spinning, everything is moving, everywhere is energy. There must be some way of availing ourselves of this energy more directly. Then, with the light obtained from the medium, with the power derived from it, with every form of energy obtained without effort, from the store forever inexhaustible, humanity will advance with giant strides. The mere contemplation of these magnificent possibilities expands our minds, strengthens our hopes and fills our hearts with supreme delight.[31]

In the audience was Robert Millikan (1868 - 1953), who won the Nobel Prize in 1923 for his work on cosmic rays. He said: 'I have done no small fraction of my research work with principles I learned that night.'[32]

THE LONDON LECTURES

Tesla then took his show to London in 1892, where he gave two lectures at the Royal Institution. There, Tesla said, James Dewar (1842 - 1923), the Institution's Professor of Chemistry, 'pushed me into a chair and poured out half a glass of a wonderful brown fluid which sparkled in all sorts of iridescent colours and tasted like nectar. "Now," he said, "you are sitting in Faraday's chair and you are enjoying the whisky he used to drink."'[33]

It was not lost on Tesla that he was lecturing on the same stage where Faraday had outlined the fundamental principles of electromagnetic induction in the 1830s. Again he put on a show of sparks, glowing wires, no-wire motors and coloured lights that spelt out the name William Thomson, the leading engineer, mathematician and physicist who was in the audience, and who that year, had become Lord Kelvin. Again he ran high-volt AC current through his body, to the amazement of his audience of distinguished scientists.

He also melted and vaporized tinfoil in a coil and produced a new type of lamp that would disintegrate zirconia and diamonds, the hardest known substances. Then he described the ruby laser which would not be built until 1960. Most of these demonstrations were brand new, not repeats of ones he had given in America.

He also demonstrated the first vacuum tube. This would later be used to amplify weak radio signals. And he concluded the lecture with speculation that improvements could be made to the transatlantic telegraph cables so that they could carry telephone calls, and the possibility of wireless transmission. The *Electrical Review* said:

> *The lecture given by Mr Tesla ... will long live in the imagination of every person ... that heard him, opening as it did, to many of them, for the first time, apparently limitless possibilities in the applications and control of electricity. Seldom has there been such a gathering of all the foremost electrical authorities of the day, on the tiptoe of expectation.*[34]

At the end, he tantalizingly informed his listeners that he had showed them 'but one-third of what he was prepared to do'. Consequently, the audience remained in their seats and he had to deliver a supplementary lecture. He then presented Lord Kelvin with one of his early experimental Tesla Coils which would be crucial in the development of wireless transmission.

Tesla went on to wow French academicians with a lecture in Paris, before heading home to Gospic where he found his mother gravely ill. She died soon after. He later wrote: 'The mother's loss grips one's head more powerfully than any other sad experience in life.'[35]

LORD KELVIN
(1824 – 1907)

Scottish engineer, mathematician and physicist, William Thomson was knighted in 1866 and made a peer in 1892 for his services to science and engineering. He helped develop the Second Law of Thermodynamics, the mathematical analysis of electricity and magnetism, the electromagnetic theory of light, the geophysical determination of the age of the Earth and the basics of hydrodynamics. His work on submarine telegraph cables helped make Britain the hub of global communications. He perfected the mariner's compass and worked out the correct value of absolute zero. The units of the absolute temperature scale are named Kelvins in tribute to him.

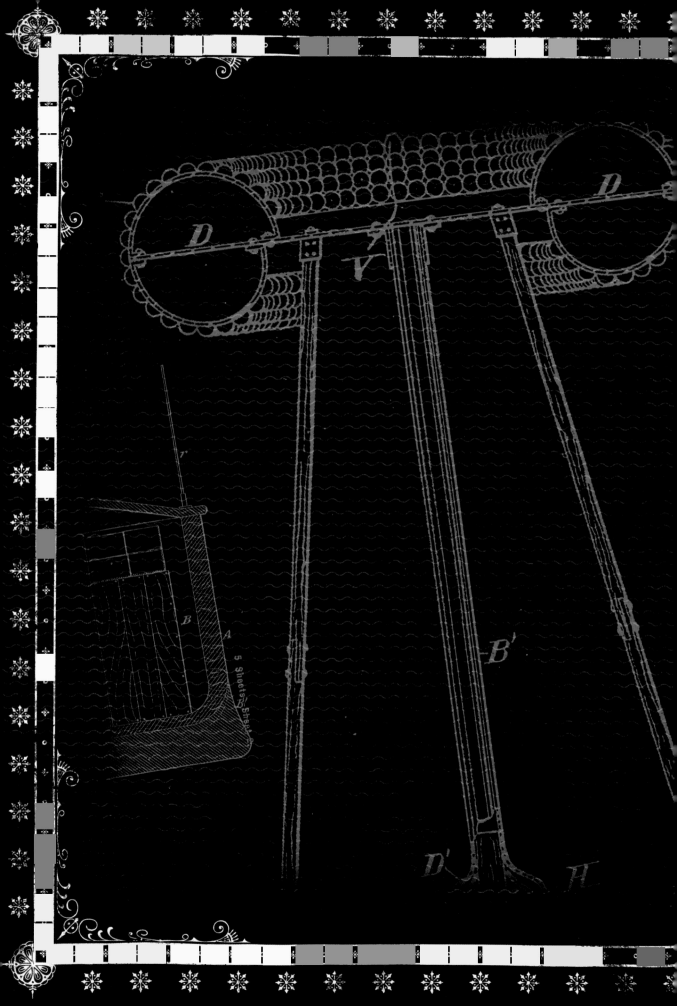

PART TWO

UNLIMITED POWERS

CHAPTER 5

FATHER OF THE WIRELESS

A mass in movement resists change of direction. So does the world oppose a new idea. It takes time to make up the minds to its value and importance. Ignorance, prejudice and inertia of the old retard its early progress. It is discredited by insincere exponents and selfish exploiters. It is attacked and condemned by its enemies. Eventually, though, all barriers are thrown down, and it spreads like fire. This will also prove true of the wireless art.

Nikola Tesla[1]

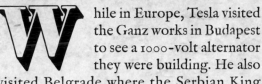
While in Europe, Tesla visited the Ganz works in Budapest to see a 1000-volt alternator they were building. He also visited Belgrade where the Serbian King Alexander I conferred the special title of Grand Officer of the Order of St Sava on him and the Serbian poet Jovan Jovanović Zmaj wrote a poem in tribute to him.[2]

On the return leg of his journey he went to Berlin to visit Hermann von Helmholtz, who developed the mathematics of electrodynamics, then went on to Bonn to see Heinrich Hertz, who was the first man to transmit and receive radio waves. Hertz had conducted his experiments with a simple sparking apparatus that could transmit radio waves across his lab. However, Hertz was a theoretical physicist who simply wanted to investigate the theories of James Clerk Maxwell, not an electrical engineer who wanted to put them to a practical use. Tesla had already duplicated Hertz's experiments and, from them, developed the Tesla Coil which was capable of transmitting wirelessly over long distances.

Hertz had sought to demonstrate that space was filled with a substance called ether, which was both inconceivably tenuous, yet extremely rigid. The reasoning was that, if light and other electromagnetic phenomena are waves, they must have something to propagate through. Tesla maintained that such a substance could not exist and the two men did not get on. In fact, the existence of Hertz's ether had already been disproved experimentally by A.A. Michelson (1852 – 1931) in Germany in 1881 and, again, in collaboration with Edward Morley (1838 – 1923) in the US in 1887.

On board ship on the way back to the US, Tesla had one of his epiphanies. He was thinking about an experience he had while walking in the Alps. Observing an oncoming thunderstorm, he noticed that rain held off until the first flashes of lightning and wondered whether he could use electricity to control the weather.

When Tesla arrived back in New York after his triumphal trip, he was greeted with a photograph of Edison signed: 'To Tesla from Edison.'[3] Then Westinghouse dropped by with the news that they had won the contract to provide the power for the forthcoming 1893 World's Fair in Chicago.

THE WIRELESS TRANSMISSION OF POWER

Hertz had conducted his experiments with a battery and a simple circuit interrupter, like a Morse key, connected to an induction coil - a small transformer - to produce a high-voltage spark. This could be detected using a copper loop with a spark gap.

Tesla quickly realized that, instead of a battery with a circuit interrupter, it would be better to use an AC current. While a circuit interrupter would only give a frequency of, at best, a few hundred cycles per second, an alternator could give 10,000 or 20,000 cycles per second. However, once an alternator reached that speed it began to fly apart, but higher frequencies could be generated electrically.

He had already used induction coils and capacitors - electrical storage devices such as a Leyden jar - to give split-phase AC currents to run his motors. These could also be used to increase the frequency even higher. Putting a connecting capacitor across the terminals of a coil produced a circuit that resonated, giving a spike in output. He called this the oscillating transformer, though other experimenters began calling it the Tesla Coil. A coil coupled to a capacitor that resonates at a specific frequency is the basis of all wireless transmission.

Refining his oscillating transformers, he earthed one terminal to the city's water main, he moved around New York detecting the electromagnetic waves generated at various frequencies. Abandoning Hertz's primitive spark gaps, he used other tuned circuits and vacuum tubes as detectors. However Tesla's aim was not to transmit an intelligible signal as we use radio waves now. His goal was the wireless transmission of power.[4]

THE SKIN EFFECT

Early in these experiments, Tesla accidentally touched a high-voltage terminal and, to his surprise, was unhurt. At high frequencies, electricity exhibits what is known as the 'skin effect'. The magnetic field created pushes the current to the outside of a conductor, so it does not run through the body, damaging the nerves and

ELECTRICAL FOREFATHERS

JAMES CLERK MAXWELL (1831–79)

Born in Scotland, James Clerk Maxwell had already demonstrated colour photography and worked on the standardization of electrical units when, in 1865, he published *A Dynamical Theory of Electrical Field*. In it, he sought to convert the physical laws of electromagnetic induction discovered by Faraday into mathematics. His famous equations showed that electric and magnetic fields travel through space as waves moving at the speed of light. This led him to propose that light was electromagnetic radiation and predicted the existence of radio waves.

HERMANN VON HELMHOLTZ (1821 – 94)

Like many scientists of the day, Helmholtz worked in multiple fields, including physiology, optics, meteorology, hydrodynamics and the philosophy of science. He is best known for the Law of Conservation of Energy. Between 1869 and 1871, he studied electrical oscillation, and he noted the oscillations of electricity in a coil when it was connected to a Leyden jar. He sought to measure the speed of electromagnetic induction, but left the determination of the length of electric waves to his star pupil, Heinrich Hertz.

HEINRICH HERTZ (1857 – 94)

After studying under Helmholtz, Hertz began his investigation of the theories of James Clerk Maxwell. He developed primitive equipment to generate electromagnetic waves and measured their wavelengths and velocity. Demonstrating that they could be reflected and refracted like light and radiant heat, he showed that light and heat were also electromagnetic waves. He was just 36 when he died.

MICHELSON-MORLEY EXPERIMENT

Devised by A.A. Michelson and later refined with Edward Morley, the Michelson-Morley experiment sought to detect the velocity of the Earth through the all-pervading ether which Helmholtz and Hertz maintained electromagnetic waves were propagated through. A sensitive interferometer was used to compare the speed of light in two directions at right angles to each other. If the universe is filled with ether, the speed of light along the Earth's direction of travel should be less than its velocity at right angles to it. No difference was detected. Ergo the ether did not exist.

muscles. Instead it travels across the surface, leaving the internal structure undamaged. In his public demonstrations, he touched one terminal of a high-frequency apparatus generating tens of thousands of volts and illuminated a bulb or tube held in the other hand. This also showed that alternating current, if at a sufficiently high frequency, was safer than direct current.

ON THE ROAD

Tesla and his lectures hit the road in 1893, and pulled in huge audiences with his dazzling demonstrations and novel ideas. In Philadelphia, he outlined a method of transmitting pictures - that is, television. The secret of wireless transmission, he said, was resonance. Wires become unnecessary as electrical impulses jump from a sending device to a receiver if they are tuned to the same frequency, and he presented a diagram showing aerials, transmitters, receivers and earth connections, all the elements of a modern broadcast system.

This was not just theory. He gave practical demonstrations. On one side of the stage he had a high-voltage transformer connected to a bank of Leyden jars, a sparking gap, a coil and a length of wire hanging from the ceiling. On the other side was an identical length of wire and an identical coil and bank of Leyden jars. But instead of the sparking gap there was a Geissler, or discharge tube that glowed when electricity was passed through it, like a primitive neon light.

Not only was the demonstration dazzling, it was full of strange sounds. When electricity was fed to the transformer, the core strained, making odd groaning sounds. Corona sizzled around the edges of the foil on the Leyden jars and sparks cracked across the sparking gap. But the radio waves travelled noisily from one antenna to the other and the Geissler tube lit up.

GUGLIELMO MARCONI (1874 – 1937)

In 1894, Marconi began experimenting with an induction coil, a Morse key and a sparking gap, along with a simple detector, at his father's estate near Bologna. Devising a simple aerial, he increased the range to 1.5 miles (2.4 km). He moved to London where he filed his first patent in June 1896. Using balloons and kites, he increased the range still further. In 1899, signals were sent across the English Channel and the America's Cup used Marconi's equipment for ship-to-shore communication. The following year Marconi took out patent No. 7777, which enabled several stations to operate on different frequencies. This was overturned by the US Supreme Court in 1943 when it was shown that Tesla and others had already developed radio-tuning circuits.

In December 1901, Marconi transmitted a signal across the Atlantic from Cornwall in England to Newfoundland in Canada. This led to the discovery that the curvature of the Earth had proved no obstacle because radio waves reflected off ionized layers in the upper reaches of the atmosphere. Marconi continued to improve the range and efficiency of wireless devices and set up companies to exploit his discoveries. In 1909 he was awarded the Nobel Prize for physics and in 1932 the Marconi company won the contract to establish short-wave communication between England and the countries of the British Empire.

Nikola Tesla quietly reads a book in his Colorado Springs laboratory in 1899, surrounded by
millions of volts of electricity and streams of artificial lightning created by a giant Tesla Coil.
The great inventor later revealed that it was a trick image produced by time-lapse photography.

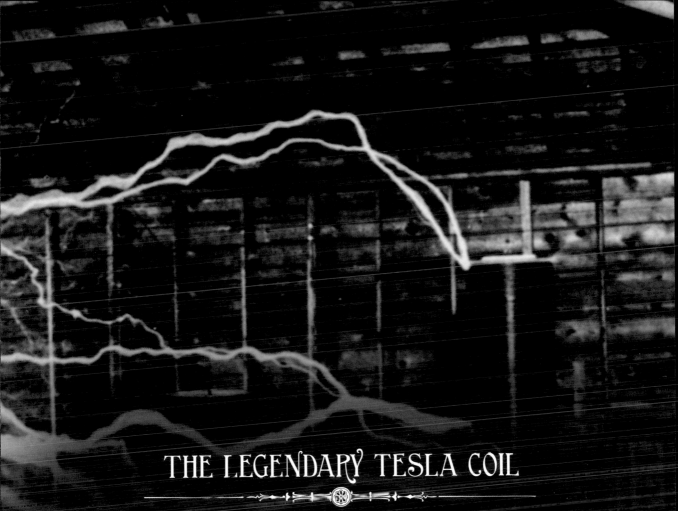

THE LEGENDARY TESLA COIL

Tesla had used coils and capacitors when experimenting with rotating magnetic fields for his motors. He continually refined the components, particularly the special transformer, or coil, at the heart of the circuit, taking out his first patent for a device to run a new and more efficient lighting system in 1891. The basic circuit connected a power supply to a large capacitor, the coil or inductor and the electrodes of an adjustable spark-gap. As the capacitor charges up, the voltage lags behind. In an inductor, voltage is felt immediately while current is held back as it has to work against the magnetic field its own passage causes.

If the size of the coil and condenser are selected to have exactly opposite timing – with voltage peaking in the capacitor just as it reaches a minimum in the coil – current and voltage can be made to chase each other back and forth. This oscillation is initiated by the spark gap. When the voltage in a capacitor builds up, it reaches a level when the air in the gap, which acts as an insulator, breaks down due to ionization and lets current flow.

In a Tesla Coil the inductor is the primary coil of a transformer. When the circuit sparks, all the energy stored over several microseconds is discharged in a powerful impulse, producing a high voltage in the secondary coil. Once the energy has been discharged, the voltage across the spark gap falls and the air becomes an insulator again, until the voltage in the capacitor builds up to the required level again. This whole process can repeat itself many thousands of times per second.

In a Tesla Coil, secondary winding is designed to react quickly to a sudden energy spike. These electrical impulses propagate along the winding as waves. The length of the coil is calculated so that, when the wave crests reach the end and are reflected back, they meet and reinforce the waves behind them, so it appears that the voltage peaks are standing still. If the high-voltage end of the secondary coil is attached to an aerial, it becomes a powerful radio transmitter. In the early decades of radio, most practicable radio transmitters used Tesla Coils. Tesla himself used larger or smaller versions of his invention to investigate fluorescence, X-rays, radio, wireless power and even the electromagnetic nature of the earth and its atmosphere.

Tesla was advised to play down the possibilities of his wireless system. It seemed so fanciful it might deter conservative businessmen who might otherwise be interested in his motors or his lighting systems. Nevertheless, he said, it earned him the title of 'Father of the Wireless'[5] among fellow researchers. Others had investigated the phenomenon of wireless transmission before him, but Tesla had pioneered the use of the tuned circuit, the aerial and the ground connection. He was giving these demonstrations of wireless transmission a full year before Guglielmo Marconi even began experimenting.

In 1896, Tesla received a letter from Sir William Preece (1834 - 1913) of the Imperial Post Office in London, asking Tesla for two wireless sets for trial. But Marconi was in London by then. He intervened, telling Preece that he had tried the Tesla system and it had not worked. Nevertheless, Tesla filed a patent for wireless transmission in September 1897.[6]

CONSPIRING WITH THE DEVIL

In St Louis, Missouri, 4,000 copies of a small-circulation electrical journal were sold because it carried an article about Tesla. When Tesla came to town, 80 electrical utility wagons paraded down the street. The 4,000-seat Grand Music Entertainment Hall was filled to overcapacity as several thousand more packed in. Tickets were being sold by scalpers for between $3 and $5 ($80 and $130 at today's prices). Tesla did not disappoint, passing 200,000 volts through his body. He described the experiment in his published lecture:

I now set the coil to work and approach the free terminal with a metallic object held in my hand, this simply to avoid burns. As I approach the metallic object to a distance of 8 or 10 inches, a torrent of furious sparks breaks forth from the end of the secondary wire, which passes through the rubber column. The sparks cease when the metal in my hand touches the wire. My arm is now traversed by a powerful electric current, vibrating at about the rate of one million times a second. All around me the electrostatic force makes itself felt, and the air molecules and particles of dust flying about are acted upon and are hammering violently against my body.

So great is this agitation of the particles, that when the lights are turned out, you may see streams of feeble light appear on some parts of my body. When such a streamer breaks out on any part of the body, it produces a sensation like the pricking of a needle. Were the potentials sufficiently high and the frequency of the vibration rather low, the skin would probably be ruptured under the tremendous strain, and the blood would rush out with great force in the form of fine spray or jet so thin as to be invisible, just as oil will when placed on the positive terminal of a Holtz machine [electrostatic generator]. The breaking through of the skin though it may seem impossible at first, would perhaps occur, by reason of the tissues under the skin being incomparably better at conducting. This, at least, appears plausible, judging from some observations.

I can make these streams of light visible to all, by touching with the metallic object one of the terminals as before, and approaching my free hand to the brass sphere, which is connected to the second terminal of the coil. As the hand is approached, the air between it and the sphere, or in the immediate neighbourhood, is more violently agitated, and you see streams of light now break forth from my fingertips and from the whole hand. Were I to approach the hand closer, powerful sparks would jump from the brass sphere to my hand, which might be injurious. The streamers offer no particular inconvenience, except that in the ends of the fingertips a burning sensation is felt ...

DEVELOPING ELECTRIC LIGHTING

The inventors of early electric lighting knew two ways to produce illumination – either by creating a spark, or arc, between electrodes or by running a current through a wire or fibre, heating it up until it glowed. Arc lamps are very bright and were used in searchlights, floodlights, lighthouses and movie projectors. They were not suitable for domestic use. But heated filaments also have their drawbacks. Most materials don't behave well when heated near their melting points. They oxidize, unless surrounded by vacuum or inert gas, or break apart through internal stress. Joseph Swan (1828 – 1914) in 1878 and Thomas Edison the following year independently developed the carbon-filament bulb. This was superseded by the more efficient tungsten-filament bulb in 1908.

However, there is another problem with incandescent lamps. In a domestic 60-watt light bulb, for example, no more than a few per cent of the energy radiated is visible. Most is lost as heat. But in 1859, Alexandre-Edmond Becquerel (1820 - 91) discovered that certain substances fluoresced when a current was passed through a Geissler tube – that is, a partially evacuated glass cylinder.

Tesla developed this into the phosphorescent lamp – phosphorescent substances are slower to emit light than fluorescent ones and continue to glow for some time after the power is switched off. He began by powering conventional filament or arc lamps with high-frequency currents. This caused the diffuse gases inside to glow and made certain solid materials give off light. The bulbs remained cold because most of the electrical energy passed through them turned into light, rather than heat. Consequently, they were much more efficient. But although he used these experiments to illustrate his celebrated lectures, he seldom patented them.

Having developed apparatus that produced higher frequencies and voltages than were available to anyone else, by 1890, he was able to light phosphorescent tubes without connecting wires. The energy was transmitted to them at radio frequencies. At higher energies, Tesla's tubes gave off X-rays.

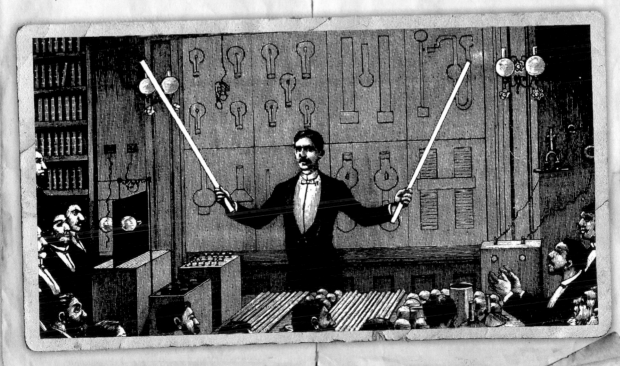

The Man with Flaming Swords. Tesla lectures before the French Physical Society and the International Society of Electricians, Paris, France, 1892.

TUNING-IN TO EARLY RADIO

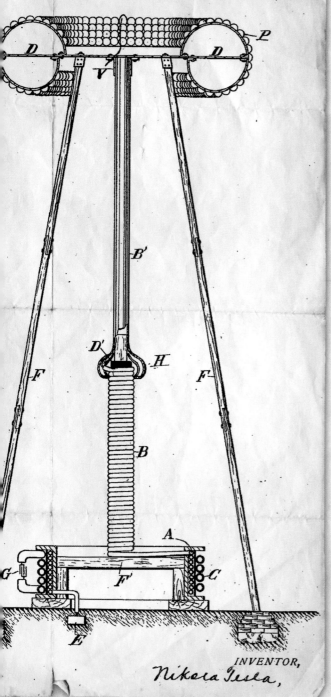

N. TESLA.
APARATUS FOR TRANSMITTING ELECTRICAL ENERGY.
LICATION FILED JAN. 18, 1902. RENEWED MAY 4, 1907.

Patented Dec. 1, 1914.

INVENTOR,
Nikola Tesla,

Tesla's patent application for an 'Apparatus for Transmitting Electrical Energy'.

Radio communication uses electromagnetic waves in the range of frequencies lying between around one hertz, or one cycle per second and a few gigahertz, or a thousand million cycles per second. The wavelength of electromagnetic radiation is the distance from one crest to the next. In an ordinary AM broadcast signal, say about 1,000 kHz on the AM dial, wave crests are spaced at about 969 ft (295 m) apart. The number of crests going by in one second is called the frequency. In all waves, the frequency multiplied by the wavelength equals the velocity. The velocity of radio waves is the speed of light.

The height of a wave is called its amplitude. In radio waves that is given in volts. For waves of the same voltage or amplitude, the more of them that arrive per second the higher the energy, so higher frequency gives more power. A single burst of high-frequency gamma rays is extremely dangerous, while the lower frequencies of radio disperse more easily and pack less of a punch.

Radio communication requires a transmitter that produces a signal powerful enough to be detected some distance away, while at the same time incorporating useful information in that signal. The receiver must be able to pick up that signal and extract the information. Once scientists had understood the nature of electromagnetic waves and detected them, the race was on to produce frequencies and voltages high enough to make wireless transmission. With what he called the 'magnifier coil' – now known as the Tesla Coil – Tesla had found a way to produce both.

It quickly became clear to Tesla that, once radio transmission had become practicable, the airwaves would be full of signals. What was needed was a way to develop circuits that worked on pre-selected frequencies so that the desired signal could be picked out from among a background of static and unrelated radio traffic. Tesla perfected tuning which lies at the heart of all communication systems today.

The streams of light which you have observed issuing from my hand are due to a potential of about 200,000 volts, alternating in rather irregular intervals, sometimes like a million times a second. A vibration of the same amplitude, but four times as fast, to maintain which over three million volts would be required, would be more than sufficient to envelop my body in a complete sheet of flame. But this flame would not burn me up; quite contrarily, the probability is, that I would not be injured in the least. Yet a hundredth part of that energy, otherwise directed; would be amply sufficient to kill a person...[7]

Waving various shaped tubes in the powerful electromagnetic field his oscillating transformer had produced, Tesla created beautiful effects like the 'spokes of a wheel of glowing moonbeams',[8] the *Electrical Engineer* said.

Towards the end of the performance, Tesla held up one of his phosphorescent lamps, the precursory of fluorescent lights, and announced that he would illuminate it by touching the terminal of his oscillating transformer with his other hand. When he did, the lamp lit up.

'There was a stampede in the two upper galleries and they all rushed out,' said Tesla. 'They thought it was some part of the devil's work.'[9]

AN ALL-AMERICAN JOKE

Returning to New York, Tesla acquired his US citizenship. To get back at Edison for his jibe years earlier of 'You are still a Parisian', Tesla decided, now that he was a fully-fledged American, to find out if Edison could take an all-American joke. He set up an experiment pitting a carbon-filament incandescent light that Edison had invented against an identical bulb that was empty.

Applying a current at a frequency of around one million cycles per second,

the empty bulb glowed brightly - more brightly than Edison's bulb which was being run on direct current. What's more, the empty bulb stayed cool to the touch. Edison was far from amused and, once again, Tesla had shot down his former mentor in the popular press.

An incandescent light bulb is only 5 per cent efficient. The other 95 per cent is lost in the form of heat. This waste, Tesla said, was 'on a par with the wanton destruction of whole forests for the sake of a few sticks of lumber'.[10]

TESLA KEEPS INVENTING

However, to Tesla, the attention of the press was a distraction. He went to work increasing the power of his oscillators until he reached one million volts. Then he immersed a high-frequency oscillator in a vat of oil. By modulating the frequency he could get the oil to rotate at different rates.

Tesla then invented a new steam-driven generator that produced as much power as one 40 times its size. Instead of using the piston action of the steam engine to turn a crankshaft and flywheel, which then turn the generator, he put the cylinder inside the coils of the generator so that the metal pistons moving up and down generated electricity.

CHICAGO WORLD'S FAIR 1893

Westinghouse had won the contract to light the World's Columbian Exposition, aka the Chicago World's Fair, in May 1893 by putting in a bid much lower than that of General Electric, which now owned Edison's patents. The buildings at the fair were to be illuminated with 200,000 bulbs, so this was an ideal opportunity to demonstrate how Tesla's AC system could be used to light an entire city.

The Electricity Pavilion at The World's Columbian Exposition in Chicago in 1893 at night.

GE had initially estimated that it would cost $1.8 million to light the fair. When this was rejected, they revised it down to $554,000. Westinghouse came in at $399,000. At that price, Westinghouse had to devise a more economical system. In less than six months, they designed and built bigger generators than had ever been built before. Using AC at high-voltage, they could distribute this throughout the fair on thin wires, saving hundreds of thousands of dollars worth of copper. The fair site would be a blaze of light and consume three times the amount of electricity then being utilized by the whole of the city of Chicago.

Westinghouse also had come up with a new design for an incandescent lamp to avoid infringing Edison's patents and manufactured 250,000 of them. Consequently when Westinghouse went to see Tesla when he arrived back in New York, he had not put much thought to promoting his motors or his polyphase system. But realizing the importance of the World's Fair as a showcase, Tesla went to Pittsburgh, he said, 'to bring the motor to high perfection'.[11]

WORDS ARE NOT ENOUGH

The Columbian Exposition covered almost 700 acres (283 Hectares) and attracted some 28 million visitors from all over the world. The centre-piece was a Ferris wheel standing 264 ft (80 m) high that could carry over 2,000 people. It revolved on the largest one-piece axle ever forged. But it was Westinghouse's illuminations that took the breath away. Former governor of Illinois, Will E. Cameron said:

Inadequate words have been found to convey a realizing idea of the beauty and grandeur of the spectacle which the Exposition offers by day, they are infinitely less capable of affording the slightest conception of the dazzling spectacle which greets

the eye of the visitor at night ... Indescribable by language are the electric fountains. One of them, called 'The Great Geyser', rises to a height to 150 ft [45 m], above a band of 'Little Geysers' ... so bewildering no eyes can find the loveliest, their vagaries of motion so entrancing no heart can keep its steady beating.[12]

VISITING THE ELECTRICITY PAVILION

At the Chicago World's Fair, the Electricity Pavilion rose to 169 ft (52 m) and covered 3.5 acres (1.4 Hectares) - the size of two soccer fields. In it, AEG exhibited the equipment they had used to transmit AC the record-breaking 109 miles (175 km) from Lauffen to Frankfurt in Germany. GE also demonstrated their new AC system. Both were technically infringing Tesla's patents, but Westinghouse made no objection as it helped demonstrate the superiority of AC. Instead, they erected a 45-ft (14 m) high monument to the 'Westinghouse Electric & Manufacturing Co. Tesla Polyphase System'.[13]

Not to be outshone, GE erected a 82-ft (25 m) Tower of Light in the centre of the Electricity Pavilion, with 18,000 bulbs around the pedestal, which was topped by a huge Edison light bulb.

Other stands showed electric body invigorators, charged belts and electricity hairbrushes. It was then thought that electricity could cure all ills. Elihu Thomson exhibited a high-frequency coil that could produce a spark 5 ft (1.5 m) long. Alexander Graham Bell launched a telephone that transmitted sound on a beam of light, while Elisha Gray (1835 - 1901) unveiled a prototype fax machine called the teleautography - for a few cents, you could have your signature reproduced electronically at a distance. Edison himself exhibited his phonograph, the multiplex telegraph and his kinetescope, which produced moving pictures for an individual viewer.

On the Westinghouse stand, Tesla exhibited AC motors and generators, and had the names of famous electrical pioneers - Franklin, Helmholtz, Faraday, Maxwell and Henry - all spelt out in phosphorescent tubes, along with that of Jovan Jovanović Zmaj (his old friend, the Serbian poet). Huge flashing neon signs saying *Westinghouse* and *Welcome Electricians* were lit by discharges of artificial lightning that made a deafening sound. Among the flashing sparks and the tubes, lit wirelessly, was a large Egg of Columbus spinning furiously.[14]

THE WIZARD OF PHYSICS

Tesla visited the World's Fair in August to put on a week of demonstrations and to attend the International Electrical Congress being held there. Its honorary chairman was Helmholtz, who Tesla showed his personal exhibit. A thousand electrical engineers attended, including most of the leaders in the field. Ten dollars were offered for seats to see Tesla, who was introduced as the 'Wizard of Physics'.[15] However, entrance was limited to those who could produce the appropriate credentials.

Tesla demonstrated mechanical oscillators and steam generators that were so small it was said they could fit in the crown of a hat. He produced motors that could run so precisely they could be used as electric clocks and a continuous-wave radio transmitter, the implications of which were lost on most of his distinguished audience. He also exhibited a version of his Egg of Columbus which demonstrated his theory of planetary motion. The *Electrical Experimenter* said:

> In this experiment one large, and several small brass balls were usually employed. When the field was energized all the balls would be set spinning, the large one remaining in the centre while the small ones revolved around it, like moons about a planet, gradually receding until they reached the outer guard and raced along the same field.

> But the demonstration which most impressed the audiences was the simultaneous operation of numerous balls, pivoted discs and other devices placed in all sorts of positions and at considerable distances from the rotating field. When the currents were turned on and the whole animated with motion, it presented an unforgettable spectacle. Mr Tesla had many vacuum bulbs in which small, light metal discs were pivotally arranged on jewels and these would spin anywhere in the hall when the iron ring was energized.[16]

The Columbian Exposition had proved to its 28 million visitors that AC was safe. From then on, over 80 per cent of all electrical devices bought in the US worked on alternating current.

TESLA'S FAMOUS FRIENDS

As a result, Tesla was proclaimed 'Our Foremost Electrician'[17] and hailed as the 'New Edison'.[18] But Tesla's health was failing again, due to overwork. Tesla's friend, Thomas Commerford Martin introduced him to socialites Robert and Katherine Underwood Johnson who took him under their wing. Tesla began calling them 'the Filipovs' after a Serbian poem, *Luka Filipov*, he had translated for them. Robert Johnson was associate editor of the prestigious *Century* magazine that ran a new profile of Tesla.

His regular dinners with the Johnsons, particularly those at Thanksgiving and Christmas, became the closest thing he knew to home life. He would arrive in a hansom cab, which would have to wait outside for hours to take him back to his hotel which was only a few blocks away.

The Johnsons were the only people with whom he was on first-name terms, except for the railroad-heir William 'Willie' K. Vanderbilt (1849 - 1920) who would let Tesla use the Vanderbilt box at the Metropolitan Opera House. Apart from opera, Tesla enjoyed theatrical comedies, particularly those featuring actress Elsie Ferguson who, he said, 'knew how to dress and was the most graceful woman he had ever seen on the stage'.[19] Gradually, he stopped going to the opera and the theatre, going to the movies instead.

It was through the Johnsons that Tesla met the writer Mark Twain, who was an admirer. Tesla told Twain that his books had saved his life when he was a boy of 12, struck down with a bout of malaria. This, apparently, brought tears to the author's eyes.[20]

Visiting Tesla's laboratory, Twain asked whether the inventor could come up with a high-frequency electrotherapy machine that he could sell to rich widows in Europe on his next visit. Tesla said he already had a machine that would aid their digestion. It vibrated in sympathy with the peristaltic waves that moved food through the gut. Enthusiastic, Twain insisted that he tried it out. It worked - too well - and sent the great writer dashing for the lavatory.

'I think I will start with the electrotherapy machine,' said Twain when he returned. 'I wouldn't want the widows to get too healthy all in one shot.'[21]

The Johnsons also introduced Tesla to the hero of the Spanish-American War, Richmond Pearson Hobson, who became a life-long friend, naturalist John Muir, who invited him out to Yosemite Valley, and writer Rudyard Kipling, who had come to live in Vermont. After dining with the author in 1901, Tesla wrote to Mrs Johnson: 'What is the matter with ink-spiller Kipling? He actually dared to invite me to dine in an obscure hotel where I would be sure to get hair and cockroaches in the soup.'

With Twain and other notables in the laboratory, the first photographs under phosphorescent light were taken. However, despite his overwork, Tesla refused to accept the Johnsons' invitation to take a holiday with them at their home at the Hamptons on Long Island.

FAME, BUT NO FORTUNE

With Tesla's help, Thomas Martin published *The Inventions, Researches and Writings of Nikola Tesla* in 1894. But Tesla kept giving copies away free. Both Martin and the Johnsons were worried that Tesla made no effort to make money out of his work and suggested that he should, at least, tell the newspapers about taking photographs under phosphorescent light so he would get the credit. Meanwhile, Martin had to lend Tesla money from his share of the book - money that Tesla would never repay.

The University of Nebraska offered Tesla an honorary doctorate, but this was considered too trifling an accolade for the great inventor. Instead, Johnson organized an honorary doctorate from Columbia. One from Yale soon followed.

To boost his fame, Martin arranged for Tesla to have his voice recorded on a phonograph, an honour already bestowed on the Australian opera singer Nellie Melba (1861 - 1931) and Sarah Bernhardt. He also got Tesla to sit for a sculptor and do interviews with the mainstream media. Journalists flocked around. Joseph Pulitzer (1847 - 1911) - who later established the Pulitzer prizes but was then publisher of the *New York World* - sent a young reporter named Arthur Brisbane (1864 - 1936) to interview Tesla in one of his favourite haunts, Delmonico's Restaurant where, for many years, he ate every night. Brisbane noted the famous restaurateur lowered his voice at the mention of Tesla's name. According to Brisbane, Charles Delmonico said in hushed tones:

> *That Tesla can do anything. We managed to make him play pool one night. He had never played, but he had watched us for a little while. He was very indignant when he found that we meant to give him 15 points. But it didn't matter much, for he beat us all even and got all the money. There are just a few of us who play for 25 cents, so it wasn't the money we cared about, but the way he studied out pool in his head, and then beat us, after we had practised for years, surprised us.*[22]

Brisbane said he found that Tesla 'stoops - most men do when they have no peacock blood in them. He lives inside of himself. He takes a profound interest in his own work.'[23] However, the engraving that accompanied the article famously showed Tesla erect and unbowed.

When asked what it was like to subject himself to such huge voltages, Tesla said: 'I admit that I was somewhat alarmed when I began these experiments, but after I understood the principles, I could proceed in an unalarmed manner.'[24]

Later he explained the spectacle presented when he was connected to an AC voltage of two-and-a-half million volts. It was, Tesla said:

> *...a sight marvellous and unforgettable. One sees the experimenter standing on a big sheet of fierce, blinding flame, his whole body enveloped in a mass of phosphorescent wriggling streamers like the tentacles of an octopus. Bundles of light stick out from his spine. As he stretches out the arms, thus forcing the electric fluid outwardly, roaring tongues of fire leap from his fingertips. Objects in his vicinity bristle with rays, emit musical notes, glow, grow hot. He is the centre of still more curious actions, which are invisible. At each throb of the electric force myriads of minute projectiles are shot off from him with such velocities as to pass through the adjoining walls. He is in turn being violently bombarded by the surrounding air and dust. He experiences sensations which are indescribable.*[25]

VOW OF CHASTITY

Tesla also became a close friend of society architect Stanford White, designer of Madison Square Garden, the Washington Memorial Arch and the New York Herald Building. They met in 1891 when piano virtuoso Ignacy Paderewski (1860 - 1941) played at Madison Square Garden for five nights. White even put Tesla up at his club, The Players, which became one of the inventor's favourite haunts. But it was a strange friendship. While Tesla was thought to be chaste, White, though married, invited him to one of his parties where naked girls emerged from pies. White was later shot dead by Harry Thaw, the jealous husband of White's mistress showgirl Evelyn Nesbit.

It seems from their correspondence that Katherine Johnson took some amorous interest in Tesla. He was also seen dining out with women. However, he had become interested in Buddhism and seems to have sworn a vow of chastity after meeting Swami Vivekananda (1863 - 1902) at a dinner with Sarah Bernhardt. Swami was in America for the Congress of World Religions held at the Chicago World's Fair and preached chastity as a path to enlightenment.

Both Bernhardt and Vivekananda visited Tesla's laboratory in New York. Tesla also studied the theosophical theories of the spiritualist Madame Blavatsky (1831 - 91), now widely seen as a charlatan.[26]

Mark Twain and Joseph (Joe) Jefferson, a well-known American actor of the time, in Tesla's South Fifth Avenue laboratory, 1894, with a blurred image of Tesla between.

TESLA'S FAMOUS FRIENDS

MARK TWAIN (1835 – 1910)

Born Samuel Langhorne Clemens, Twain was an American humorist and writer who found worldwide fame with *The Adventures of Tom Sawyer* (1876) and *The Adventures of Huckleberry Finn* (1885). He was also known for his travel writing – *The Innocents Abroad* (1869), *Roughing It* (1872) and *Life on the Mississippi* (1883).

JOSEPH RUDYARD KIPLING (1865 – 1936)

Born in Bombay (Mumbai), India, Kipling was a short-story writer, poet and novelist who chronicled the British Empire at the height of its power. He also wrote for children. Principally remembered for the adventure novel *Kim*, *The Jungle Book*, *Just So Stories*, the short-story *The Man Who Would be King* and the poems *Mandalay*, *Gunga Din*, *The White Man's Burden* and *If–*, he won the Nobel Prize for Literature in 1907.

JOHN MUIR (1838 – 1914)

Born in Scotland, Muir emigrated to the US with his family in 1849. After studying science at the University of Wisconsin, he found work in a factory where he adapted and improved machinery. An accident nearly cost him his sight. In its aftermath he undertook a walk of nearly 1,000 miles (1,600 km) from Indiana to Florida. In 1868, he arrived in the Yosemite Valley in California and became an advocate for the preservation of the wilderness there. Due to his lobbying, National Parks were set up at Yosemite, Sequoia and elsewhere. In 1903, he accompanied President Theodore Roosevelt on a camping trip in the Yosemite region.

RICHMOND PEARSON HOBSON (1870 – 1937)

Graduating from the US Naval Academy in 1889, Hobson was given temporary command of the collier *Merrimac* during the Spanish-American War. Off Cuba, his ship was disabled by enemy fire and he scuttled her in the entrance to Santiago Harbour, blockading the Spanish Fleet. He and his crew of six were imprisoned in Morro Castle. When he was released in a prisoner exchange in 1898, he returned to the US to a hero's welcome. Women admirers flocked to him and he became 'the most kissed man in America'. Awarded the Medal of Honor, he became a congressman. One of Tesla's closest friends, he said the inventor once told him that he had 'never touched a woman'.

CHAPTER 6

NIAGARA FALLS

In the schoolroom there were a few mechanical models which interested me and turned my attention to water turbines. I constructed many of these and found great pleasure in operating them. How extraordinary was my life an incident may illustrate. My uncle had no use for this kind of pastime and more than once rebuked me. I was fascinated by a description of Niagara Falls I had perused, and pictured in my imagination a big wheel run by the Falls. I told my uncle that I would go to America and carry out this scheme. Thirty years later I saw my ideas carried out at Niagara and marvelled at the unfathomable mystery of the mind.

Nikola Tesla[1]

In 1886, civil engineer Thomas Evershed, who had worked on the Erie Canal, proposed digging a series of canals and tunnels to carry water from Niagara Falls to waterwheels that would be used to power industrial mills and factories. Three years later, Edison drew up a plan to electrify the city of Buffalo, NY, which was 20 miles (32 km) away. However, DC had never been transmitted more than one or two miles.

Even Westinghouse, at that time, was dubious that electricity could be transmitted so far and suggested a complex system of compressed air tubes and cables to convey the power. Plans were drawn up for the construction of an industrial complex next to the Falls, but then came the news that AC power had been transmitted the 109 miles (175 km) from Lauffen to Frankfurt by AEG in Germany.

The International Niagara Commission, headed by Lord Kelvin, offered $20,000 for the best plan to harness the power of the Falls. Like Edison, Kelvin was opposed to AC - until he saw it in action at the Columbian Exposition. Then he became an enthusiastic convert. Westinghouse refused to enter at first as he felt that, to win, he would be handing over $100,000-worth of advice. Of the twenty schemes submitted, fourteen used hydraulics or compressed air.[2] Four involved DC power, one of which was endorsed by Edison. Two used AC. One of them was not fully worked out; the other used the Tesla system manufactured by Westinghouse.

CLOSING THE DEAL

GE thought they were still in the running and when blueprints went missing from the Westinghouse works they were accused of industrial espionage. However the success of the hydroelectric plant at Telluride followed by Westinghouse's triumph at the Chicago World's Fair left no one in doubt about who should be awarded the contract. Thomas Martin's article on Tesla in *Century* closed the deal. The following year, *The New York Times* wrote: 'To Tesla belongs the undisputed honor of being the man whose work made this Niagara enterprise possible ... There could be no better evidence of the practical qualities of his inventive genius.'[3]

Meanwhile the president of the Cataract Construction Company, Edward Dean Adams, visited Tesla in New York and offered him $100,000 for a controlling interest in fourteen US and foreign patents, along with any future inventions Tesla may come up with. Tesla accepted and in February 1895 the Nikola Tesla Company was set up. Not only was Tesla working on wireless and remote control, he was putting his mind to cheap refrigeration, the production of liquid air, the manufacture of fertilizers and nitric acid from the air, and artificial intelligence.

ELECTRIFYING BUFFALO

Construction of the first power station at Niagara took 5 years. It was a headache for investors throughout. The outlay was huge and no one knew whether it would work as the plans lay principally in Tesla's three-dimensional imagination. However their worries evaporated when the switch was thrown and the first power reached Buffalo at midnight on 16 November 1896. The *Niagara Gazette* reported: 'The turning of a switch in the big powerhouse at Niagara completed a circuit which caused the Niagara River to flow uphill.'[4] The first 1,000 horsepower of electricity reaching Buffalo was taken by the street railway company, but already the local power company had orders from residents for 5,000 more. Within a few years the number of AC generators at Niagara Falls

reached the planned ten, and power lines ran as far as New York City. Broadway was ablaze with lights. It powered streetcars and the subway system. Even Thomas Edison's networks converted to alternating current.

MESMERIZED BY MARS

While these developments were going on, Tesla was doing more experiments with wireless transmission. He set up a transmitter on the roof of his laboratory and using an aerial strung from a balloon, he could detect a signal on top of the Hotel Gerlach, thirty blocks away.

As always, Tesla was a visionary. Walking up Fifth Avenue one fine Sunday afternoon in 1894, he said to his young assistant D. McFarlan Moore: 'After we have signalled from any point to any point on the Earth, the next step we will be signalling other planets.'[5]

America was in the grip of Martian fever at the time. The noted astronomer Percival Lowell (1855 - 1916) was studying the 'canals' on Mars and John Jacob Astor (1864 - 1912) - the richest man to die on the *Titanic* - had just published *A Journey to Other Worlds*. He gave a copy to Tesla.

For the time being, Tesla was planning to see if he could receive signals from a ferry on the Hudson River, but on 13 March 1895 his laboratory burnt down. While Tesla was wrestling with depression, Westinghouse was fighting over the patents for Tesla's AC induction motors against GE and others. GE, of course, promulgated the theory that the fire at Tesla's lab had been caused by the sparks emanating from one of his motors. In fact, it had started on the floor below.

Tesla set about finding a new lab. In the meantime, Edison let him use a workshop in Llewellyn Park, New Jersey, and, although uninsured, Tesla was confident that Westinghouse would pay for any new equipment he needed.

However, Westinghouse was a hard-headed businessman and billed Tesla. Meanwhile, he announced that he was planning to use Tesla's motors, whose patents he owned, to power locomotives.

The following year, 1896, Tesla told the press that he was looking into the 'possibility of beckoning Martians'[6] and, when Lord Kelvin arrived in America in 1897, he suggested using the lights of New York to flash a signal to the Martians. Meanwhile Edison was working on something even more outlandish - a telephone to contact the dead.

But for Tesla contacting Mars was just an 'extreme application of [my] principle of propagation of electric waves'.[7] It was merely an extension of a more Earthly goal. He pointed out: 'The same principle may be employed with good effects for the transmission of news to all parts of the Earth ... Every city on the globe could be on an immense circuit ... a message sent from New York would be in England, Africa and Australia in an instant. What a grand thing that would be.'[8]

ELECTRIC DEMON DUO

Arthur Brisbane in *The World* newspaper had announced that Tesla was 'greater even than Edison',[9] but New York's *Troy Press* asked: 'Who is electric king, Edison or Tesla?'[10] Meanwhile, the two men, now billed as the 'Twin Wizards of Electricity',[11] were appearing at the National Electrical Exposition in Philadelphia. Tesla was then on the ascendant as AC had been transmitted along telephone lines for a record-breaking 500 miles (800 km). Tesla was disappointed though as the power at the Exposition was restricted due to the fear of fire.

By this time, Edison was conceding: 'The most amazing thing about this Exposition is the demonstration of the ability to deliver here an electric current

Left: A giant turbine at Niagara Falls hydroelectric power station.

generated at Niagara Falls. To my mind it solves one of the most important questions associated with electrical development.' Bell concurred, stating, 'This long distance transmission of electric power was the most important discovery of electric science that had been made for many years.'[12]

Tesla told the *Philadelphia Press*: 'I am now convinced beyond any question that it is possible to transmit electricity ... to commercial advantage over a distance of 500 miles at half the cost of generation by steam ... I am willing to stake my reputation and my life on this declaration.'[13]

THE POWER OF ELECTRICAL HEALING

Following on from Mark Twain's idea, Tesla began to experiment on the healing properties of electricity in his new laboratory on Houston Street in Greenwich Village. At the time, doctors were promoting electricity as a 'vitality booster' and a 'universal healing agent'.[14] Some even said that it could cure tuberculosis, which was rife at the time. It was reported that Tesla took daily doses to deal with his depression after his lab burnt down. He said that high-frequencies 'produce an anti-germicidal action'. As part of his daily routine, he would strip off and climb on board his apparatus and crank up the juice.[15]

He was also said to be working on an electric weeding tool to clear railroad tracks of unwanted undergrowth. He paid a short visit to Colorado, where he claimed to have transmitted a signal through Pike's Peak, using the energy of the Earth, rather than his oscillators. Announcing the success of this experiment in Arthur Brisbane's newspaper *The World* on 8 March 1896, he said: 'Electricity would be as free as the air. The end has come to telegraph, telephone companies, and other monopolies ... with a crash.'[16]

X-RAYS, SHADOWGRAPHS AND COSMIC RAYS

While running a current through partially evacuated glass tubes, Tesla had also noticed a special radiation was given off that could be detected by phosphorescent and fluorescent substances. In 1892, he gave lectures on what he called 'black light and very special radiation'. Experimenting with his radiation he notice that he could produce what he called 'Shadowgraphs' on plates inside metal containers. Unfortunately, these were lost when his laboratory burned down.

When he read of Röntgen's discovery of X-rays, he realized that these were the same thing as his 'very special radiation'. He produced more shadowgraphs and sent them to Röntgen who asked how they had been made.

Tesla quickly realized that he could get better results with a Tesla Coil that developed 4 million volts.[17] While others were X-raying thin structures such as hands and feet, he was taking photographs through the skull at a distance of 40 ft (12 m) from the tube.

While experimenting, Tesla noticed that the energy had both particle and wave-like attributes, something later recognized by Albert Einstein.[18] He also speculated that the tiny lumps of matter involved, later known as electrons, might be broken up into even smaller pieces and said that 'similar streams must be emitted by the Sun'[19] - what we now know as cosmic rays.

Tesla X-rayed birds and animals, himself and his assistants, quite oblivious to the fact that this might be dangerous. He reasoned that the amount of material involved was so small that it would take centuries to build up enough to be poisonous. He himself suffered from bad headaches when experimenting with X-rays and an assistant suffered blistering and inflammation of exposed skin.[20]

Edison was also experimenting with X-rays and noted that they caused sensations in the eyes of the blind. He believed that eyesight might be restored by the application of X-rays. Tesla disagreed and there was another falling out.[21]

The rift was mended when the Kentucky School of Medicine combined devices made by Tesla and Edison to remove birdshot from the foot of a voter who had been shot during an election scuffle. Thomas Martin then took Tesla, Edison and other electricians on a fishing trip off Sandy Hook. Ironically, Tesla caught a large flounder; Edison a huge fluke.[22]

HEADING FOR THE FALLS

In July 1896, Tesla, Westinghouse, Adams and others involved in the Niagara project, travelled up to the Falls. On their arrival, the *Niagara Gazette* reported:

> *Tesla is an idealist, and anyone who has created an ideal of him from the fame that he has won will not be disappointed in seeing him for the first time. He is fully six feet tall, very dark of complexion, nervous, and wiry. Impressionable maidens would fall in love with him at first sight, but he has no time to think of impressionable maidens. In fact, he has given as his opinion that inventors should never marry. Day and night he is working away at some deep problems that fascinate him, and anyone that talks with him for only a few minutes will get the impression that science is his only mistress, and that he cares more for her than for money and fame.*[23]

Tesla was overcome at the sight of the Falls and the first of the hydroelectric power stations designed by Stanford White built there. It would house some ten gigantic Tesla turbines generating over 35,000 kilowatts.

Afterwards he returned to New York City, where he threw himself back into research into the wireless transmission, fearing that Marconi may steal a march. Again he refused to holiday with the Johnsons, though he did have Christmas dinner with them.

RESPECT, ACCLAIM AND KUDOS

The celebration for the inauguration of the Niagara Falls power station was held in January 1897 in the Ellicott Club in Buffalo, NY. The top 350 of America's most prominent businessmen made the trek there. A notable no-show was Thomas Edison.

Tesla was introduced as the 'greatest electrician on Earth' and received a standing ovation. However, Tesla made a rambling, self-deprecating speech, saying it had been a mistake to invite him. He heaped praise on those who had helped. Running out of time, the master of ceremonies intervened and cut off the end of his speech.[24] Just as well, as a blissfully unaware Tesla was about to enlighten the distinguished audience by telling them that they had wasted all their time and money building a power line from Niagara to Buffalo - he would soon be transmitting the electricity *wirelessly* ...[25]

His continual self-deprecation did him no favours. Others were claiming to have invented the induction motor and the Tesla Coil, and they were pirating his inventions. Meanwhile he turned down several applications to be his assistant from a top Yale student, Lee De Forest (1873 - 1961), who eventually went on to rival Marconi in the development of radio.[26]

CHAPTER 7

TESLA'S EXTREME SCIENCE

Suppose the whole earth to be like a hollow rubber ball filled with water, and at one place I have a tube attached to this, with a plunger in the tube. If I press upon the plunger the water in the tube will be driven into the rubber ball, and as the water is practically incompressible, every part of the surface of the ball will be expanded. If I withdraw the plunger, the water follows it and every part of the ball will contract. Now, if I pierce the surface of the ball several times and set tubes and plungers at each place the plungers in these will vibrate up and down in answer to every movement which I may produce in the plunger of the first tube. If I were to produce an explosion in the centre of the body of water in the ball, this would set up a series of vibrations in the whole body. If I could then set the plunger in one of the tubes to vibrating in consonance … in a little while and with the use of a very little energy I could burst the whole thing asunder.

Nikola Tesla, explaining a global telegraph system[1]

Back in New York, Tesla began developing Elisha Gray's teleautography into telephotography. Edison then announced that he planned to launch the autographic telegraph, which would allow journalists to file their stories effortlessly, along with sketches and pictures. Tesla claimed his system could also work wirelessly, at a time when sending a Morse signal still had to be perfected.

Tesla had studied a system developed in 1846 by Scottish physicist Alexander Bain (1810 - 77). It transmitted pictures using a grid of wires imbedded in wax under a sheet of chemically treated paper. The receiver used the same grid where an electric stylus drew the shape. Tesla found that it was better to break down the elements of the picture using one wire and a spinning disc. Dr Arthur Korn of the University of Munich, who transmitted a photograph in 1902, cited his debt to Tesla.[2] These experiments were the basis of the fax machine and the television.

CONNECTING TO THE EARTH'S ENERGY

From what he read, Tesla began to suspect that Marconi was using clones of Tesla's equipment in his experiments. After Sir William Preece had cancelled the test of Tesla's equipment, Lloyds of London contacted Tesla and asked if he would rig up a ship-to-shore system for an international yacht race in 1896. Tesla high-handedly refused, fearing that his work would be confused with the amateurish efforts of others in the field.[3]

He then began secret experiments that he did not even tell his lab assistants about. He would set up his transmitter in East Houston Street, then take a battery-powered receiver up the Hudson River to West Point, a distance of some 50 miles.

From there, he could tune in to the signal from the transmitter. He did this two or three times, he told a court in 1915.

At the same time, he considered harnessing wind power, tidal power, solar energy and geothermal energy. Electricity could be used to electrolyze water, separating it into hydrogen and oxygen, whose explosive recombination would produce heat and steam. He patented a machine to produce ozone and worked out how to separate nitrogen out of air electrically. The farmer would simply shovel earth into the machine and switch it on. The current would drive out the oxygen and hydrogen, leaving the nitrogen to be absorbed in the soil which would emerge ready-fertilized.

Over 4,000 people turned out to see his lecture on the advances he had made in the field of X-rays at the New York Academy of Science, though it is thought that they had hoped to see him hurling thunderbolts again.[4] Then in an article in *Scribner's Magazine* on Marconi's successful transmission of a radio signal 8 miles (13 km), he outlined a system for transmitting messages instantly around the world using the telluric currents that run below the surface of the Earth. He also had plans to transmit signals through the ionized layers thought to exist in the upper atmosphere.[5]

While Tesla had done all the early development in radio, Marconi was preparing to transmit a signal across the English Channel. Once again Tesla had failed to exploit his own invention. Without the money to pursue his bigger projects, his pronouncements made him sound like a mad scientist. Brown and Peck were still earning thousands from his patents, while Westinghouse had joined forces with GE. Tesla's induction motors and polyphase system were about to power subway trains without a penny going to the inventor.

Tesla was further sidelined at an electrical exhibition in New York organized by

Stanford White. The Marconi company was represented by Edison's son, Tom Edison Jr. Marconi had needed some wireless patents that Edison had taken out, and the Wizard of Menlo Park was happy to do business.[6]

MAKING THE EARTH MOVE

Tesla placed one of his oscillators in the central support beam in the basement of the building of his Houston Street lab and adjusted the frequency until the beam began to hum. While he was distracted momentarily, the building began to shake, along with the earth and all the buildings around it. According to the *Brooklyn Eagle*: 'The Fire Department responded to an alarm frantically turned in; four tons of machinery flew across the basement and the only thing which saved the building from utter collapse was the quick action of Dr Tesla in seizing a sledgehammer and destroying his machine.'[7]

Tesla called the device a 'Frankenstein's monster', and pointed out that no building could stand the strokes of a 5-pound hammer, delivered at its resonant frequency. On another occasion, Tesla claimed to have gone down to Wall Street where there was a ten-storey steel frame of a building under construction, clamped an oscillator the size of an alarm-clock to it and tuned it in.

In a few minutes, I could feel the beam trembling. Gradually the trembling increased in intensity and extended throughout the whole great mass of steel. Finally, the structure began to creak and weave, and the steel workers came to the ground panic-stricken, believing that there had been an earthquake. Rumours spread that the building was about to fall, and the police reserves were called out. Before anything serious happened, I took off the vibrator, put it in my pocket, and went away. But if I had kept on 10 minutes more, I could have laid that

building flat in the street. And, with the same vibrator, I could drop the Brooklyn Bridge in less than an hour.[8]

He told reporters that he could have split the Earth the same way, destroying mankind. He had worked out that the resonant frequency of the Earth has a periodicity of 1 hour, 49 minutes. If he were to explode a ton of dynamite every 1 hour, 49 minutes, the shock waves would keep reinforcing one another. He estimated that it would take a year to smash the world to pieces, 'but in a few weeks I could set the earth's crust into such a state of vibration that it would rise and fall hundreds of feet, throwing rivers out of their beds, wrecking buildings and practically destroying civilization. The principle cannot fail.'[9]

RESEARCHING REMOTE CONTROL

In 1898, the United States went to war with Spain after the battleship, the USS *Maine*, was sunk in Havana harbour. Cuba was still a Spanish colony at the time. It was thought that Tesla was on John Jacob Astor's yacht when, to aid the war effort, he proposed the idea of a guided torpedo.

While Astor and his yacht went to war, Tesla began making preliminary experiments with a remote-controlled boat. Tesla had a large tank in the auditorium of the Electrical Exhibition in 1898. In it was a 4 ft (1.2 m) boat. By means of transmitters working at various frequencies, he could start and stop the boat, steer it and switch its lights on and off.[10] He had also planned to build a submersible, perhaps to stage mock battles between Spanish ships and the American fleet. But he was upstaged by the Marconi company who were demonstrating remotely controlled mines, detonated wirelessly. The press got particularly excited when Tom Edison Jr accidentally blew up his desk where other mines were stored.[11]

CREW-LESS DEVIL AUTOMATA

Tesla's invention seemed all the more crazy when he proposed a *Torpedo Boat Without a Crew*:

> My submarine boat, loaded with its torpedoes, can start out from a protected bay or be dropped over a ship's side, make its devious way below the surface, through dangerous channels of mine beds, into protected harbours and attack a fleet at anchor, or go out to sea and circle about, watching for its prey, then dart upon it at a favourable moment, rush up to within a hundred feet if need be, discharge its deadly weapon and return to the hand that sent it. Yet through all these wonderful evolutions it will be under the absolute and instant control of a distant human hand on a far-off headland, or on a war ship whose hull is below the horizon and invisible to the enemy.
>
> I am aware that this sounds almost incredible and I have refrained from making this invention public till I had worked out every practical detail of it. In my laboratory I now have such a model, and my plans and description at the Patent Office at Washington show the full specifications of it.[12]

Even the *Electrical Engineer*, edited by his friend Thomas Martin, complained that Tesla was always promising great things and failing to deliver, saying: 'Mr Tesla fools himself, if he fools anybody, when he launches forth into the dazzling theories and speculations associated with his name.'[13]

> He would tether up aloft balloons in those strata and deliver to them large quantities of current at such high potential that it would travel economically across the space without wires, say from Niagara Falls to Paris. By this facile distribution of water power, coal and steam would become unnecessary to industry. The new plan may explain why Mr Tesla has abandoned his old steam oscillator. It is earnestly to be hoped that this novel idea will prove workable. Balloons were a dismal failure in our late war, but that is no criterion, and Mr Tesla may have some superior gas for inflation and sustentation purposes. It will be remembered that Mr Marconi has already telegraphed from balloon to balloon, without wires, a distance of over 20 miles, thus proving in advance the tenability of Mr Tesla's proposition.[14]

Perhaps Tesla was caught up in war fever but he was convinced his 'Devil Automata' was the way of the future. 'The continuous development in this direction must ultimately make war a mere contest of machines without men and without loss of life,' he wrote, 'a condition which would have been impossible without this new departure, and which, in my opinion, must be reached as preliminary to permanent peace.'[15]

Others agreed with him. One of them was Mark Twain who wanted to sell patents to European governments. Tesla himself entered into negotiations with Czar Nicholas II (1868 - 1918) of Russia. Nevertheless, some began to write him off. The journal *Public Opinion* compared his remote-control boat to the mysterious 'motive power' of John Worrell Keely (1837 -98) who had just died, and said: 'The facts of Mr Tesla's inventions are few and simple as the fancies which have been woven around it are many and extravagant. The principle of the invention are not new, nor was Tesla the original discoverer.'[16]

BEAUTIFUL BUT INCOMPLETE INVENTIONS

Tesla was upset by Thomas Martin's attack and wrote a response that *Electrical Engineer* was forced to publish. It said: 'Being a bearer of high honours from a number of American universities, it is my duty, in view of this slur,

THE REMOTE CONTROL BOAT

Tesla's tub-like craft was powered by large batteries on board. Radio signals activated switches, which controlled the boat's propeller, rudder and running lights. But even registering the arrival of a radio signal pulse taxed the rudimentary technology of 1898. Tesla had to invent a new kind of coherer or a radio-activated switch to achieve this. The coherer was a canister with a little metal oxide powder in it. The powder orients itself in the presence of an electromagnetic field, such as radio waves, and becomes conductive. In Tesla's coherer, the canister flips over after the signal has passed, restoring the powder to a random, non-conductive state.

Each signal advanced a disc one step, making a new set of contacts. So if the contacts had previously given the combination 'right rudder/propeller forward full/light off', the next step might combine 'rudder center/propeller stop/lights on'. The connected circuits operated levers, gears, springs and motors, then flipped the coherer over so it was ready to receive the next instruction. Tesla assumed that this system could be used on radio-guided torpedoes. But it was too far ahead of its time. The US Navy did finance some trials in 1916, but the money went to one of Tesla's competitors as his patent had expired. Nevertheless, as the 20th century progressed, more and more uses were found for remote control.

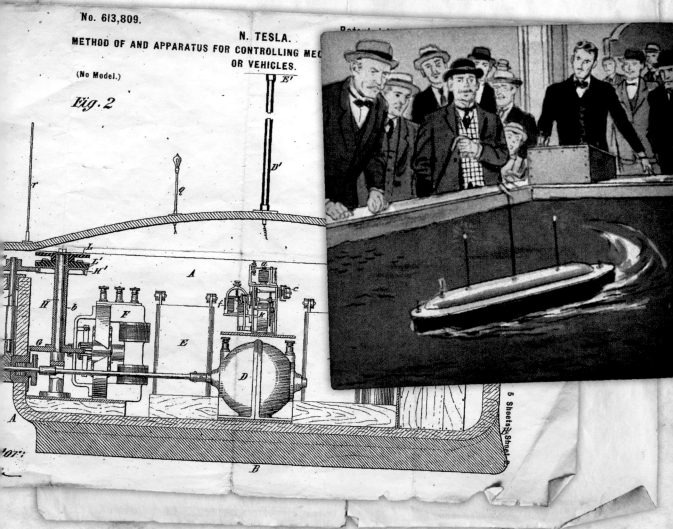

Left: Tesla Patent - Apparatus for Controlling Mechanism of Moving Vessels or Vehicles.

Right: Tesla demonstrates a radio-controlled boat to the public during an electrical exhibition at Madison Square Garden, 1898.

to exact from you a complete and humble apology ... On this condition I will again forgive you, but I would advise you to limit yourself in your future attacks to statements for which you are not liable to be punished by law.'[17]

Martin struck back, saying directly after Tesla's letter: 'Our foremost electrical inventor has been kind enough to say that the *Electrical Engineer* made Mr Tesla.'[18] The implication, of course, being that Tesla was not America's foremost electrical engineer. And it was true that between 1890 and 1898, *Electrical Engineer* had published 167 articles by or about Tesla. In that time, *Electrical Review* had published 127 and *Electrical World* just 97.[19]

The rebuttal was headed 'His Friends to Mr Telsa' and urged him to complete a long list of 'beautiful but unfinished inventions', but he should stop making statements about such fantastic things as remote-control aircraft that would 'explode at will... [and] never make a miss'.[20] The world was not ready at that point in time for the cruise missile that Tesla was describing.

But Martin had a point. Tesla's oscillator was not a commercial success. His fluorescent tubes never went on the market. And his wireless transmission of power was never realized. Tesla was sanguine. He wrote later: 'I'm glad that I am living in a place in which, though they can roast me in the papers, they cannot burn me at the stake.'

As a final shot at Martin and the *Electrical Engineer*, he published an article in *Electrical Review* with pictures showing him, holding a glowing wireless vacuum lamp the size of a basketball lit by millions of volts conducted by his body.[21]

I, ROBOT ...

A reporter from *The New York Times* watching his remote control boat said he could envision a wireless torpedo. Tesla had a bigger vision. 'You do not see there a wireless torpedo,' he said, 'you see there the first of a race of robots, mechanical men which will do the laborious work of the human race.' For Tesla it was a short leap from a remote-controlled machine to one that could think for itself. In *Century* magazine in June 1900, he wrote: 'I am an automaton endowed with power of movement, which merely responds to external stimuli beating upon my sense organs, and thinks and acts and moves accordingly. I remember only one or two cases in all my life in which I was unable to locate the first impression which prompted a movement or a thought, or even a dream.'[22] Consequently, a sentient being could be manufactured.

> *Long ago, I conceived the idea of constructing an automaton which would mechanically represent me, and which would respond, as I do myself, but, of course, in a much more primitive manner, to external influences. Such an automaton evidently has to have motive power, organs for locomotion, directive organs, and one or more sensitive organs so adapted as to be excited by external stimuli ... Whether the automaton be of flesh and bone, or of wood and steel, it matters little, provided it can perform all the duties required of it like an intelligent being.*[23]

However, people found it hard to take his ideas seriously. Tesla called his remotely controlled boat 'The First Telautomaton', but the examiner-in-chief of patents found the concept so unbelievable that he had to come and see it for himself. And when he thought of offering it to the government, the official in Washington he spoke to burst out laughing.

During the Spanish-American war, the Secretary of the Navy also turned down Tesla's offer of wireless transmitters to help coordinate ship and troop movements for fear of the sparks that they might give off. Tesla assured him that he had overcome this problem, but the persistent image of Tesla with lighting bolts pouring from his fingers was too vivid.

IGNITED BY COSMIC FORCES

Tesla also believed that we are shaped by cosmic forces 'not in the vague and delusive sense of astrology, but in the rigid and positive meaning of physical science'. After all, science 'admits that the suns, planets, and moons of a constellation are one body, and there can be no doubt that it will be experimentally confirmed in times to come, when our means and methods for investigating psychical and other states and phenomena shall have been brought to great perfection'.[24]

The spark of life was present in every inanimate object too. 'Even matter called inorganic, believed to be dead, responds to irritants and gives unmistakable evidence of a living principle within,' he said. 'Thus, everything that exists, organic or inorganic, animated or inert, is susceptible to stimulus from the outside ... What is it that causes inorganic matter to run into organic forms? ... It is the Sun's heat and light. Wherever they are there is life.'[25]

NOT MAD AT ALL

Some people still had faith. Tesla boasted that he had produced a lamp that was far superior to the incandescent bulb, using one-third of the energy.

As my lamps will last forever, the cost of maintenance will be minute. The cost of copper, which in the old system is a most important item, is in mine reduced to a mere trifle, for I can run on a wire sufficient for one incandescent lamp more than a thousand of my own lamps, giving fully five thousand times as much light.

On the strength of this, Tesla's friend John Jacob Astor invested $100,000 in the Tesla Electric Company and Tesla moved into the Waldorf-Astoria.

HAMMOND AND SON

Mining engineer and philanthropist John Hays Hammond (1855 – 1936) gave Tesla $10,000 to develop his Telautomaton. Later his son, John (Jack) Hays Hammond Jr (1888 – 1965), developed Tesla's ideas and became known as 'The Father of Remote Control'. At Yale, Jack developed electrically controlled steering and engine control for a boat, controlling the mechanisms at a distance using a wireless device. In 1909, he got his father to arrange a meeting for him with Tesla because there was some 'important information' he needed from him. With Marconi wireless, which itself used Tesla Coils, attached to two 360-ft (110 m) towers, Jack could control a crew-less boat from a lookout station near his laboratory at Freshwater Cove. Later, Jack invited Tesla to speak at his graduation from Yale.

CHAPTER 8

IN COLORADO SPRINGS

Nikola Tesla, the Serbian scientist, whose electrical discoveries are not of one nation, but the pride of the world, has taken up his abode in Colorado Springs ... On East Pike's Peak avenue, with limitless plains stretching to the eastward, and a panorama of mighty mountains sweeping away north and south, to the west – Tesla has caused to be constructed a [wireless] station for scientific research.

Desire Stanton, Colorado Springs, 1899[1]

With Tesla's coils now generating up to 4 million volts with sparks jumping from the walls to the ceilings, Tesla's Houston Street laboratory was becoming a fire hazard. Nor was it secure against the snooping of Edison's spies. And Tesla had experiments that he wanted to conduct, he said, in secret.

He had been out to Pike's Peak outside Colorado Springs in 1896 at the invitation of Westinghouse patent attorney Leonard E. Curtis. For his new experiments, he needed huge amounts of power, but he would be working mostly late at night when the load would be least and Curtis arranged for him to get free power from the local utility, the El Paso Power Company.

After stopping to show off his Telautomaton in Chicago, he arrived in Colorado Springs on 18 May 1899 and immediately breached his own secrecy. When a reporter asked him what his plans were, he said: 'I propose to send a message from Pike's Peak to Paris, France. I see no reason why I should keep the thing a secret any longer.'[2] He was welcomed with a banquet. Mining camps in the area had adopted his AC system, so he was already a celebrity out West.

With a local carpenter named Joseph Dozier, he built an experimental station on an empty field known as Knob Hill, which had a view over Pike's Peak to the west and rolling plains to the east. It was essentially a wooden barn measuring 60 ft by 70 ft (18 m by 21 m). It consisted of one large room with a roof that opened, two small offices at the front and a balcony.

Again intent on keeping the exact nature of his experiments secret, Tesla had the only window that Dozier had provided, boarded up. A fence ringed the station with numerous signs on it saying: KEEP OUT, GREAT DANGER. Above the door was a phrase from Dante's *Inferno* said to be the inscription above the entrance to Hell: *Abandon hope, all ye who enter here.*[3]

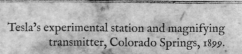

Tesla's experimental station and magnifying transmitter, Colorado Springs, 1899.

THE MAGNIFYING TRANSMITTER

Colorado Springs was 6,000 ft (1,800 m) above sea level and Tesla planned to tap into the rarefied air 5,000 ft (1,500 m) above the Earth. He soon discovered that the 10 ft (3 m) helium balloons he had bought from Germany could not lift the hundreds of feet of wire, so he devised a telescopic mast that raised a copper ball to a height of 142 ft (43 m). To steady the mast, Tesla built a 25 ft (8 m) tower on the roof of his laboratory.

Under it, he built a 'magnifying transmitter'. On top of a 6 ft (1.8 m) wall, he laid two turns of thick cable. This was fed 500 volts from the end of a streetcar line that stopped just short of Knob Hill. Current was passed through a 50-kilowatt Westinghouse transformer, stepping the voltage up to 20,000 or 40,000 volts.[4]

In the centre of the room, was a secondary coil comprising hundreds of turns of finer wire. One end was connected to a round terminal inside the laboratory or the copper ball on top of the mast. The other end was earthed. The apparatus was completed by a bank of capacitors that could be discharged by a motorized brake-wheel, while other large coils moved in and out of the magnetic field.

Tesla began experimenting with wireless telephones, reporting to Astor: 'There is nothing novel about telephoning without wires to a distance of 5 or 6 miles [8 or 9 km], since this has been done often before ... In this connection, I have obtained two patents.'[5]

TAKING THE PULSE OF THE PLANET

Tesla also experimented sending electrical signals through the earth. Then on 4 July 1899 a huge electrical storm arrived. He recorded 'no less than 10,000 to 12,000 discharges being witnessed inside of 2 hours.

The flashing was almost continuous and, even later in the night, when the storm had abated, some 15 to 20 discharges per minute were witnessed. Some of the discharges were of a wonderful brilliancy and showed often ten or twice as many branches.'[6]

He could track these discharges with his sensitive detecting equipment and he noted that they registered, periodically, even when the storm had moved out of sight. They seemed to start and stop every half an hour. Tesla concluded that the lightning strikes had created electromagnetic waves in the earth's crust which, reflected back on themselves, set up stationary waves. These moved past the receiver as the storm receded.[7]

While Marconi could send radio waves across the English Channel, Tesla believed that by harnessing these waves 'not only would it be practicable to send telegraphic messages to any distance without wires, but also to impress upon the entire globe the faint modulations of the human voice, far more still, to transmit power, in unlimited amounts, to any terrestrial distance without loss'.[8]

'With these stupendous possibilities in sight,' wrote Tesla, 'I attacked vigorously the development of my magnifying transmitter, now, however, not so much with the original intention of producing one of great power, as with the object of learning how to construct the best one.'[9]

MARS ON THE HORIZON

Improving his instruments, Tesla found he could detect electrical disturbances 1,100 miles (1,770 km) away. The detector was a 'coherer' - a glass tube filled with iron fillings - connected via a capacitor to the ground. This was placed within the secondary coil of the magnifying transmitter. When a signal was applied to the electrode, the iron fillings would align, allowing current in a secondary circuit to pass through it.

GREAT DANGER
KEEP OUT.

Nikola Tesla peeps out the door of the Colorado Springs
experimental station, early summer 1899.

Nikola Tesla holding a gas-filled phosphor coated wireless light bulb which he developed in the 1890s, half a century before fluorescent lamps came into use. Published on the cover of the *Electrical Experimenter* in 1919.

Tesla connected a telephone receiver across the coherer which would beep each time a signal was detected. Alone in the laboratory one night, he was surprised to hear regular beeps - first one, then two, then three.

'My first observations positively terrified me,' said Tesla, 'as there was present in them something mysterious, not to say supernatural ... I felt as though I were present at the birth of a new knowledge or the revelation of a greater truth.'[10]

He quickly discounted that a signal with 'such a clear suggestion of number and order' could have come from disturbances in the Sun, the aurora borealis or currents in the Earth. They could not be entirely accidental and the thought flashed through his mind that they might be under intelligent control. He could not decipher them, but over the next year 'the feeling was constantly growing on me that I have been the first to hear the greeting of one planet to another'.[11]

At Christmas 1900, the local Red Cross Society asked him what the greatest achievement of the next hundred years would be. In his answer, he said: 'Brethren! We have a message from another world, unknown and remote. It reads: one... two... three...'[12]

In interviews, Tesla maintained only that the signals were of an extraterrestrial origin, but journalists quickly concluded they were from Mars.[13] Thanks to Percival Lowell, everyone thought that Mars was inhabited. In his book *Mars* (1895), Lowell concluded that there had been a drought on Mars and the Martians had built canals to carry water from the polar ice caps.

Biographer Marc Seifer concluded that Tesla had picked up signals from the tests Marconi was doing with the British and French fleets.[14] But, at that time, the transmitters Marconi was using would not have had the power to transmit a signal half way round the world. Moreover, Marconi was using the high frequencies needed to carry radio waves through the air, while Tesla's equipment was tuned to the very low frequencies that he believed were transmitted better through the Earth's crust.

Another theory was advanced by Kenneth and James Corum who pointed out that Io, a moon of Jupiter, emits a signal in the 10 kHz range used by Tesla. In 1996, they built a Tesla receiver and recorded a series of bleeps like those Tesla described in 1899. Studying astronomical charts they also discovered that both Jupiter and Mars would have been in the night sky over Colorado in the summer of 1899. What's more, on several nights in July, the signal from Io would have broken off just as Mars was setting. If Tesla had walked out of his lab when the beeping stopped, he would have seen Mars disappearing over the horizon.[15]

The press began to speculate how Tesla would reply to the Martians, while others dismissed Tesla as a man who would do anything to attract self-publicity.[16]

WONDERFUL WHITE LIGHTNING

Tesla continued experimenting with his magnifying transmitter, boosting the voltage until it produced streams of artificial lightning 16 ft (5 m) long which set fire to the building more than once. Tesla was continually finding himself close to danger.

For handling the heavy currents, I had a special switch. It was hard to pull, and I had a spring arranged so that I could just touch the handle and it would snap in. I sent one of my assistants down town and was experimenting alone. I threw up the switch and went behind the coil to examine something. While I was there the switch snapped in, when suddenly the whole room was filled with streamers, and I had no way of getting out. I tried to break through the window but in vain as I had no tools, and there was nothing else to do than to throw myself on my stomach and pass under. The primary carried 500,000 volts, and I had to crawl through the narrow place ... with the streamers going. The nitrous acid was so strong I could hardly breathe. These streamers rapidly oxidize nitrogen because of their enormous surface, which makes up for what they lack in intensity. When I came to the narrow space they closed on my back. I got away and barely managed to open the switch when the building began to burn. I grabbed a fire extinguisher and succeeded in smothering the fire.[17]

Frightening though this experience was, Tesla was thrilled. 'I have had wonderful experiences here,' he wrote, 'among other things, I tamed a wild cat and am nothing but a mass of bleeding scratches. But in the scratches, there lies a mind.'[18]

Although the famous pictures of Tesla show him surrounded by lightning, they were not discharged during the normal running of the machine. They would have been a waste of energy. When the magnifying transmitter was run at night, a blue beam would be seen rising straight up over the station into the night sky, caused by a corona of fine streamers surrounding the mast and sphere.

'At night,' Tesla said, 'this antenna, when I turned on to the full current, was marvellous sight.'[19]

MISSING THE BOAT WITH THE NAVY

While in Colorado, Tesla was contacted by the US Lighthouse Board who wanted to install a wireless on board the Nantucket Lightship so that it could give advanced warning of incoming ships to New York and other east-coast ports. Tesla initially agreed to supply some experimental

Another time-lapse image from Colorado Springs, 1899. A 12 million volt discharge from the Tesla Coil creates huge artificial lightning bolts while Tesla sits reading, apparently unperturbed.

equipment to test on the lightship. But his relations with the Lighthouse Board quickly soured when he discovered that his equipment was going to be tested against Marconi's. The Lighthouse Board then said that they would prefer 'home to foreign talent'. Tesla was furious, insisting that he was the pioneer who had laid down the principles of wireless telegraphy and was not prepared to compete with the upstart Marconi.[20]

But Marconi was arriving in New York for the America's Cup in September 1899 and something must be done quickly. Tesla was adamant. He had important work to do in Colorado and would not break off unless the board put in an order for at least 12 wireless sets. They refused. So Tesla missed an opportunity to demonstrate the effectiveness of his equipment and, when the US Navy bought its first wireless equipment, it was ordered from French and German companies.[21]

TURNING UP THE JUICE

As Tesla continued to increase the power of his magnifying transmitter, he became a danger to all around him. In the article 'Can Radio Ignite Balloons?' he said:

Referring to electrical or radio wave action at a distance, I know from experience that if proper precautions are not taken, fires of all kinds and explosions can be produced by wireless transmitters. In my experiments in Colorado, when the plant was powerfully excited, the lightning arresters for 12 miles around were bridged with continuous arcs, much stronger and more persistent than those which ordinarily took place during an electric storm. I have excited loops (coil aerials) and lighted incandescent lamps at a considerable distance from the laboratory without even using more than 5 or 10 per cent of the capacity of the transmitter. When the oscillator was excited to

about 4 million volts and an incandescent lamp was held in the hand about 50 or 60 feet from the laboratory, the filament was often broken by the vibration set up, giving some idea of the magnitude of the electromotive forces generated in the space. The accompanying illustration shows one of my experiments in which I lighted several lamps at a distance of 100 feet [30 m] from the laboratory, purely by wireless energy. Such induced currents might easily fire a gas balloon under the proper conditions.[22]

Arriving for work one day in mid-autumn, his assistant, Kolman Czito, found Tesla the inventor watering the ground around the metal plate he had buried near the lab as an earth. 'If I could only insulate these wires with liquid oxygen,' he said, 'I could reduce losses another magnitude.'

He gave Czito a pair of rubber soled shoes and put on a pair himself. 'All the way today, sir?' asked Czito. 'To the limit, my friend,' said Tesla. 'Now remember, keep one hand behind your back at all times.' This was to prevent a circuit being created between his two arms that would send a lethal current through the heart.

'When I give you the signal I want you to close the switch and leave it closed until I give you the signal to open it,' said Tesla. Usually, he told Czito to close the switch for a second, no longer. Then Tesla tottered out of the lab on his high shoes, past the testing equipment and cold lamps planted in the ground at various places, and positioned himself on a knoll about a mile away where he could see the top of the mast. Even though insulated, sparks jumped from the ground to his feet as he crunched along the path.[23]

TOTAL BLACK OUT

It was already evening and the lights were going on in Colorado Springs as Tesla gave the signal and Czito closed the switch. There was a crackling sound from the coils.

The room was filled with an eerie blue light. The coils and the building itself sprouted needles of flames as the place filled with the smell of ozone.

A low rumble built to a roar of thunder that was so strong it could be heard 15 miles (24 km) away. Butterflies circled as if caught in a whirlpool and a horse half-a-mile away bolted. 'I suppose the capacity of the body was sufficiently great to derive a rather strong current through the legs which would frighten the animal,' said Tesla.

Streamers of lightning shrouded the mast high above the roof of the lab. Suddenly there was silence. Below, Colorado Springs had been plunged into darkness. Tesla raced back to the lab to berate Czito 'Why did you do that?' he shouted. 'I did not tell you to open the switch. Close it again immediately.'

Czito pointed to the meters on the switchboard. The wires carrying power to the lab were dead. When they phoned the power station, Tesla grabbed the phone.

'This is Nikola Tesla,' he said. 'You have cut off my power! You must give me back my power immediately! You must not cut off my power.'

The power station worker on the other end of the phone explained that Tesla had short-circuited their generator and had totally wrecked the power station. The generator was on fire, but fortunately the powerhouse had a second, standby generator which was started up soon after. Tesla insisted that he be supplied with current from the reserve generator as soon as it was up and running. This was refused. In the future, he was told, he would be supplied by a generator operating independently from the one supplying the El Paso Electric Company's other customers. The generator would be the one Tesla had just burnt out. It was up to him to fix it.[24]

Tesla left Colorado Springs - without paying his electricity bill - convinced that his experiments had been a success.[25]

WHAT HAPPENED IN COLORADO SPRINGS?

Tesla conducted experiments at Colorado Springs for nine months and though he kept a daily diary, it is not clear what the results of his experiments were. He had promised to transmit a wireless signal from Pike's Peak to Paris. He also aimed to transmit electricity without wires at high altitudes where the air was thinner and more conductive. He did succeed in lighting up the sky.

He transmitted extra-low-frequency signals from the surface of the earth to the ionosphere. Tesla then calculated that the resonant frequency of this area was approximately 8 Hz. It was not until the 1950s that this idea was taken seriously and researchers were surprised to discover that the resonant frequency of this space was indeed around 8 Hz.

He also found the earth to be 'literally alive with electrical vibrations'[26] and came to believe that lightning striking the ground set up powerful waves that moved from one side of the earth to the other. If the Earth was a great conductor, Tesla thought that he could transmit unlimited amounts of power through it with virtually no loss. To test this theory, he had attempted to become the first man to create electrical effects on the scale of lightning. And there were some reports that he transmitted a signal powerful enough to illuminate vacuum tubes planted in the ground several miles away. But this may be attributed to conductive properties of the ground locally at Colorado Springs.[27]

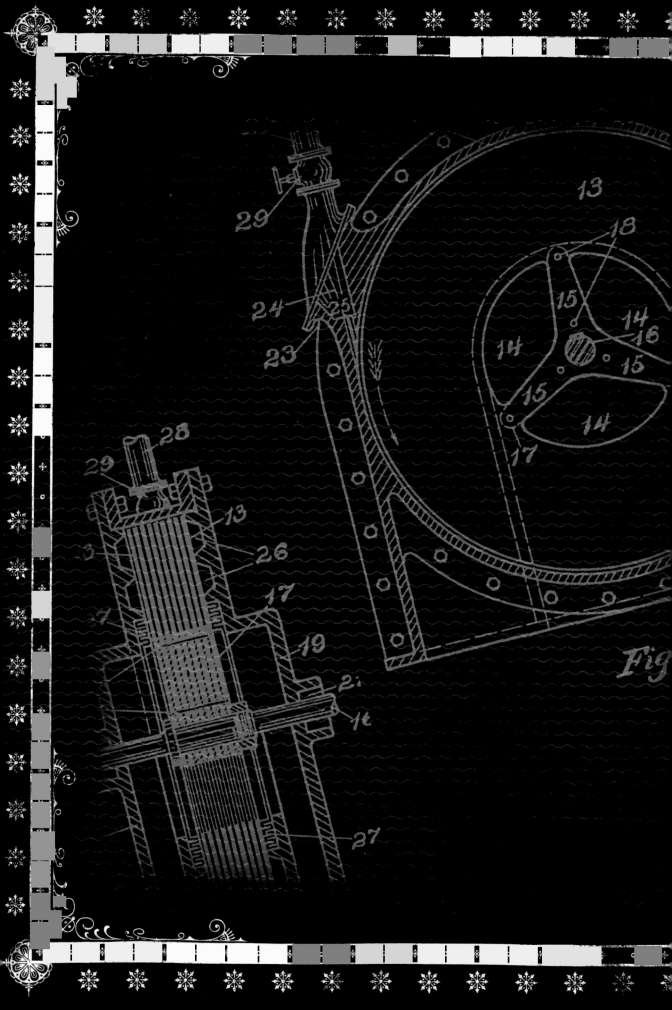

PART THREE

IMPOSSIBLE IDEAS

CHAPTER 9

TAKING ON MARCONI

When I sent electrical waves from my laboratory in Colorado around the world, Mr Marconi was experimenting with my apparatus unsuccessfully at sea. Afterward, Mr Marconi came to America to lecture on the subject, stating that it was he who sent those signals around the globe. I went to hear him, and when he learned that I was present he became sick, postponed the lecture, and up to the present time has not delivered it.

Nikola Tesla[1]

arconi was in town when Tesla returned to New York to the comfort of the Waldorf-Astoria. He paid him a visit. 'I remember him when he was coming to me asking me to explain the function of my transformer for transmission of power to great distances,' said Tesla. 'Mr Marconi said, after all my explanations of the application of my principle, that it is impossible.'

'Time will tell, Mr Marconi,' Tesla replied.[2] Tesla was still trying to get interest in his Telautomatons. He proposed a 'dirigible wireless torpedo' or small remote-controlled airships.

'I have constructed such machines, and shown them in operation on frequent occasions,' said Tesla. 'They have worked perfectly and everybody who saw them was amazed at their performance.'[3]

He went to Washington DC, where he was again rebuffed.[4] He still hoped that the US Navy or Coast Guard might buy his wireless transmitters, and planned to prove his system by transmitting a signal across the Atlantic. Westinghouse, though now in financial difficulties, fronted the money and an agent was sent to Britain to find a suitable site for a receiving station.

In 1900, Tesla filed three patents on wireless communication and reworked his plans for a transoceanic broadcasting system. He also set to work on an article for *Century* magazine. At the time he was under the influence of the philosophers Friedrich Nietzsche (1844 - 1900) and Arthur Schopenhauer (1788 - 1860). Robert Johnson begged him not to make the article too metaphysical. Tesla took no notice.

LIFE ON MARS

As well as his scientific interests the article covered such arcane subjects as the evolution of the race, artificial intelligence, the possibility of future human beings living without eating, inorganic life forms and life on Mars. However, the magazine decided they had no alternative but to publish it anyway. It was a sensation, among his friends, at least. However, anonymous reviews in *Popular Science Monthly* and *Science* dismissed it as 'science and fiction'.[5]

Oblivious to criticism, Tesla followed up with an article in *Collier's* magazine called *Talking With the Planets* where he discusses the possibility of communicating with Martians.[6] This brought renewed criticism, especially from those who had axes to grind. Meanwhile Tesla tried to get fresh funds out of Astor. They were not forthcoming. Astor was angry with Tesla. Instead of using the money he had given him before to perfect his fluorescent tube, Tesla had run off to Colorado Springs and spent it on his wireless experiments.

There were other problems on the emotional front. The son of his first - and, probably, only - love Anna turned up in New York, saying he wanted to be a boxer. Tesla, a boxing fan, encouraged him. Stanford White set up a bout, but the boy was knocked down and died soon after. Tesla, it was said, grieved for him as if he were his own son.[7]

FINDING A NEW BACKER

Tesla moved in high circles. In the autumn of 1900, he was invited to attend the wedding of Louisa, the daughter of Wall Street magnet J. Pierpont 'J.P.' Morgan (1837 - 1913), along with the Astors and Teddy Roosevelt (1858 - 1919) who became US president the following year. He turned out with a top hat, white gloves and cane. Tesla always prided himself on his appearance and claimed to be one of the best-dressed men on Fifth Avenue. At the wedding, Louisa's younger sister Anne took a fancy to Tesla. She invited him to the Thanksgiving dinner at the Morgans' home where Tesla put on a show involving coloured lights, lightning and various wireless devices.[8]

Morgan was a yachtsman and commodore of the New York Yacht Club. During the America's Cup, he offered Marconi $200,000 for his American patents, including the 'Ocean Rights ... if ever wireless telegraphy could communicate from England to New York'. But the deal fell through and Morgan opened negotiations with Tesla.

However, when they met, Morgan was not impressed. Already mired in controversy, Tesla was boastful and, aside from his early deal with Westinghouse, he had yet to show a profit on any of his inventions. Nevertheless Morgan agreed to give Tesla $150,000[10] to build a transatlantic transmitter with a 90-ft (27-m) tower in return for 51 per cent of the company and the patents. The lighting patents that Astor had an interest in were added to this later. Tesla took the opportunity to pay Westinghouse back the money he owed him for re-equipping his laboratory[11] and the new venture was celebrated with a banquet at the Waldorf-Astoria.[12]

TESLA ON BOXING

Tesla claimed to have made a study of heavyweight title fights after the 1892 match where street-fighter John L. Sullivan (1858 – 1918), who had held the world title for 10 years, was knocked out by college-educated 'Gentleman Jim' Corbett (1866 – 1933). In 1927, he made headlines predicting the outcome of the rematch between Gene Tunney (1897 – 1978) and the 'Manassa Mauler' Jack Dempsey (1895 – 1983) who, though he had lost the title a year earlier, was ahead in the betting.

The *New York Herald Tribune* said: 'Sitting in his suite at the Hotel Pennsylvania, the 71-year-old inventor did not hedge or pussyfoot, but declared that Tunney was 'at least 10 to 1 favourite'. On the basis of mechanics, Tesla said, 'Tunney will hit Dempsey continuously and at will'. He added that Tunney also had the advantage because he was single. 'Other things being equal,' Tesla said, 'the single man can always excel the married man.'[9] In his later years, Tesla would be seen dining with other boxers, including the 'Midland Mauler' Jimmy Adamick and Yugoslav welterweight Fritzie Zivic.

WORLD TELEGRAPHY CENTRE

Tesla bought a 200-acre (81-hectare) tract of land at Shoreham on Long Island Sound. It was named Wardenclyffe, after James S. Warden, the businessman who handled the deal.[13] The plan was to expand from there an 1800-acre (728-hectare) 'Radio City' which Stanford White set about designing. Travelling out to the site by train one morning, Tesla read an article in *Electrical Review* where Marconi admitted using a Tesla Coil in his wireless experiments.[14]

Enraged, Tesla immediately scrapped plans to build a modest 90-ft (27-m) tower and started designing a 600-ft (183-m) edifice. When he grew tired of commuting from the city, Tesla would stay out on Stanford White's estate near Shoreham. When White's wife asked Tesla why he wandered around the garden at night, Tesla replied: 'I never sleep.'

When Morgan got wind of Tesla's grandiose plans he fulminated. White quickly scaled things down.[15]

MARCONI'S MIRACLE

Tesla's wireless ambitions were about to suffer another setback. Marconi had installed a power transmitter with a 200-foot (60 m) mast at Poldhu in Cornwall and was sending test transmissions to Crookhaven in Ireland, 200 miles (320 km) away, while the sister station was being built on Cape Cod, MA. Both were flattened by storms in September 1901.

The Poldhu station was rebuilt, but the aerial on the other side of the Atlantic was to be carried aloft by a kite from Signal Hill, Newfoundland. On 12 December 1901, it picked up a signal - three dots, the Morse code for the letter S, from Poldhu. The age of global communication had begun.

At first no one believed it. According to Tesla, Marconi had once told him that, because of the curvature of the Earth 'wireless communication across the Atlantic was impossible because there was a wall of water several miles high between the two continents, which the rays could not traverse'.[16]

On hearing the news, Otis Pond, an engineer then working for Tesla, said, 'Looks as if Marconi got the jump on you.' Tesla replied, 'Marconi is a good fellow. Let him continue. He is using seventeen of my patents.'[17]

Thomas Martin was doubtful about Marconi's achievement, but after consulting with colleagues, he booked a banqueting hall in the Waldorf-Astoria for 13 January 1902 and invited 300 guests to celebrate Marconi's achievement. Tesla did not attend, having ducked out of the hotel before Marconi arrived. The dinner exacerbated the ill-will between Martin and Tesla, who was now preparing a new edition of *The Inventions, Researches and Writings of Nikola Tesla*, this time without Martin's name on it.

According to *The New York Times* at the dinner: There were cheers when the toastmaster came to a letter from Nikola Tesla, who said he felt that 'he could not rise to the occasion'. The letter went on:

Marconi's radio station at Poldhu, on the coast of Cornwall, England.

I regret not being able to contribute to the pleasure of the evening, but I wish to join the members in heartily congratulating Mr Marconi on his brilliant results. He is a splendid worker, full of rare and subtle energies. May he prove to be one of those whose powers increase and whose mind feelers reach out farther with advancing years for the good of the world and honour of his country.[18]

In his speech, Marconi pointed out that his wireless was already installed on over 70 ships - 37 in the British Royal Navy, 12 in the Italian Navy and the rest on liners belonging to Cunard, North German Lloyd and others. There were already 20 stations in operation on land in Great Britain and more in construction. He concluded by saying: 'I have built very largely on the work of others ... I may miss a few of them, but I would like to mention Clerk Maxwell, Lord Kelvin, Professor Henry and Professor Hertz.'[19]

There was no mention of Tesla, but he was not cowed. He wrote to Morgan saying that he had developed a machine that would produce 'an electrical disturbance of sufficient intensity to be perceptible over the whole of the Earth ... when I throw a switch, I shall send a greeting to the whole world and for this great triumph I shall ever be grateful to you.'[20]

Not only would he take on the telegraph companies, but he threatened to destroy newspapers. Every customer of Tesla's wireless system would be able to print their own. However, Tesla was already running out of money and this would be his last communication with Morgan for 9 months.[21]

THE WARDENCLYFFE TOWER

Though Tesla's experimental station was impressive, it was not as grand as he had planned. Morgan had only given him $150,000 when a more realistic sum would have been $1 million. The tower rose just 187 ft (57 m) in the air. On the top was a 57-ton steel sphere. Under the tower was a shaft that plunged 120 ft (36 m) into the ground. Sixteen iron pipes were driven down another 300 ft (91 m) so that currents could pass deep into the Earth. 'In this system that I have invented,' Tesla said, 'it is necessary for the machine to get a grip of the Earth, otherwise it cannot shake the Earth. It has to have a grip ... so that the whole of this globe can quiver.'[22]

While Marconi was sending his messages through curved air, said the *Port Jefferson Echo*, Tesla proposed to send them through the Earth as well.[23] But construction was already grinding to a halt when Marconi sent his transatlantic signal. He may have been using Tesla's patents, but his equipment was inexpensive by comparison. Then the stock market crashed and the cost of materials Tesla needed to complete Wardenclyffe doubled. Morgan would not stump up any more money, so Tesla went back to the manufacture of his oscillators that had proved so popular among his rivals, and to perfecting fluorescent lights. Money began to trickle in and he completed the cupola crowning the tower at Wardenclyffe.

At the end of July 1903, Tesla finally cranked up his magnifying transformer. The mushroom-shaped cupola became fully charged. Local villagers heard the rumble of thunder and a strange light appeared above Tesla's tower. This could be seen on the shores of Connecticut on the other side of Long Island Sound.

Soon after, creditors from Westinghouse came to cart away the heavy equipment and Tesla's tower fell silent. Unbowed, Tesla raised money from his uncles in the Balkans, then did the rounds of Wall Street financiers. In October 1903, Thomas Fortune Ryan (1851 - 1928) promised $100,000, but Morgan scuttled the venture.[24] It seems that Tesla's boast of being able to transmit unlimited amounts of power over any

Right: An image of the way Tesla's Wardenclyffe Tower would look when completed.

WORLD WIRELESS

RETURN
TO
NIKOLA TESLA CO.
8 West 40 St., N.Y.

distance was seen as a threat to the moguls of Wall Street. How were they going to charge for the electricity it generated? Tesla clearly intended to give power away for free. Even distributing information freely was a challenge to those who controlled major corporations. Tesla responded with an article published simultaneously in *Electrical World* and *Scientific American*:

> *The results attained by me have made my scheme of intelligence transmission, for which the name of 'World Telegraphy' has been suggested, easily realizable. It constitutes a radical and fruitful departure from what has been done heretofore ... It involves the employment of a number of plants, all of which are capable of transmitting individualized signals to the uttermost confines of the earth. Each of them will be preferably located near some important centre of civilization and the news it receives through any channel will be flashed to all points of the globe. A cheap and simple device, which might be carried in one's pocket, may then be set up somewhere on sea or land, and it will record the world's news or such special messages as may be intended for it. Thus the entire earth will be converted into a huge brain, as it were, capable of response in every one of its parts. Since a single plant of but one hundred horsepower can operate hundreds of millions of instruments, the system will have a virtually infinite working capacity, and it would immensely facilitate and cheapen the transmission of intelligence. The first of these central plants would have been already completed had it not been for unforeseen delays...*[25]

He said elsewhere that, if only Morgan would fund it, he would bring about world peace. But Morgan was adamant and Tesla sought refuge in Wardenclyffe, only venturing out to attend the funeral of Stanford White which, due to the scandalous circumstances, was shunned by most other New York socialites.[26]

MORE RIVALS EMERGE

Marconi was not Tesla's only rival. Lee De Forest had now completed his doctorate in

J.P. MORGAN (1837 – 1913)

The son of a successful financier, Junius Spencer Morgan (1813-90), John Pierpont Morgan began his career in 1857 with the New York banking firm of Duncan, Sherman and Company, which was the US representative of the London firm George Peabody and Company. By 1871 he was a partner at Drexel, Morgan and Company, soon the predominant source of government financing. In 1895, it became J.P. Morgan and Company, and one of the most powerful banking houses in the world. Because of his links with Peabody, Morgan was able to provide the rapidly growing US industrial corporations with capital from British banks.

Investing in railroads, by 1902, he controlled some 5,000 miles (8,000 km) of track. In 1891, he arranged the merger of Edison General Electric and Thomson-Houston Electric Company to form General Electric. In the depression that followed the panic of 1893, he formed a syndicate to resupply the US government's depleted gold reserve. Having financed the creation of the Federal Steel Company in 1898, he merged it with the giant Carnegie Steel Company in 1901 to form US Steel Corporation. The following year, he formed the International Harvester Company and the International Mercantile Marine, which dominated transatlantic shipping. He led the attempt to avert a general financial collapse following the stock market panic of 1907. Then he began amassing banks and insurance companies. This gave him control over the nation's leading corporations and financial institutions.

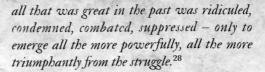
electrical engineering at Yale. In 1901, he sent wireless messages across the Hudson River. He speeded up transmission to 30 words a minute, which was about as fast as a Morse-code operator could send them. By 1904, he could send a signal from Buffalo to Cleveland, a distance of 180 miles (290 km). Then in 1908, he succeeded in bridging the Atlantic.[27]

While Marconi and De Forest concentrated on sending messages in Morse code, sending brief bursts of signals, Canadian-born Reginald Fessenden realized that it was possible to modulate a continuous signal to follow the irregularities of sound. At the receiving station, it would then be possible to unscramble the signal and reconvert it to sound. This is what we now know as AM (amplitude modulated) radio. In 1906, he transmitted music down the Massachusetts coast. In 1910, De Forest was broadcasting the voice of Italian opera singer Enrico Caruso (1873 - 1921) from the Metropolitan Opera House in New York. However, Tesla sued him for infringing his patents and won.

THE MILLION-DOLLAR FOLLY

The newspapers began to call Wardenclyffe *Tesla's Million-Dollar Folly*. Forced to close it down, Tesla had a nervous breakdown. He railed against the critics:

> *It is not a dream. It is a simple feat of scientific electrical engineering, only expensive – blind, faint-hearted, doubting world ... Humanity is not yet sufficiently advanced to be willingly led by the discoverer's keen searching sense. But who knows? Perhaps it is better in this present world of ours that a revolutionary idea or invention instead of being helped and patted, be hampered and ill-treated in its adolescence – by want of means, by selfish interest, pedantry, stupidity and ignorance; that it be attacked and stifled; that it pass through bitter trials and tribulations, through the heartless strife of commercial existence. So do we get our light. So*

> *all that was great in the past was ridiculed, condemned, combated, suppressed – only to emerge all the more powerfully, all the more triumphantly from the struggle.*[28]

Tesla retreated to his room at the Waldorf-Astoria where he nursed an injured pigeon he had found near the New York Public Library. However, at night, he sometimes stole out to Wardenclyffe to hook himself up to the high-frequency machinery. 'I have passed 150,000 volts through my head,' he told *The New York Times*, 'and did not lose consciousness, but I invariably fell into a lethargic sleep sometime after.'[29] He found the electricity soothing.

SEEING INTO THE FUTURE

Although Wardenclyffe did not live up to his expectations, Tesla's vision remained intact. In 'The Future of the Wireless Art' in *Wireless Telegraphy and Telephony* in 1908 he said:

> *As soon as it is completed, it will be possible for a business man in New York to dictate instructions, and have them instantly appear in type at his office in London or elsewhere. He will be able to call up, from his desk, and talk to any telephone subscriber on the globe, without any change whatever in the existing equipment. An inexpensive instrument, not bigger than a watch, will enable its bearer to hear anywhere, on sea or land, music or song, the speech of a political leader, the address of an eminent man of science, or the sermon of an eloquent clergyman, delivered in some other place, however distant. In the same manner any picture, character, drawing, or print can be transferred from one to another place. Millions of such instruments can be operated from but one plant of this kind. More important than all of this, however, will be the transmission of power, without wires, which will be shown on a scale large enough to carry conviction.*[30]

LEE DE FOREST (1873 – 1961)

Like Tesla, De Forest was the son of a church minister who hoped his son would follow him into the ministry. Lee spent much of his youth at Talladega College, traditionally an African-American school where his father was president. In 1893, he enrolled at the Sheffield Scientific at Yale where he studied engineering. Six years later he was awarded a PhD for a thesis entitled *Reflections of Hertzian Waves from the Ends of Parallel Wires*.

Experimenting in radio-telegraphy, he managed to interest the US Army and Navy in his apparatus. His equipment was used by European reporters to send despatches during the Russo-Japanese War of 1904 – 05. In 1906, De Forest filed a patent for a vacuum tube diode to detect radio waves. The following year, he patented the triode or Audion valve. This placed a grid between the electrodes which allowed it to amplify feeble electric currents. While others developed its full potential, it was the mainstay of amplification until the invention of the transistor. In 1912, De Forest was indicted, and subsequently acquitted, of mail fraud by seeking to promote this 'worthless device'. His triode made transcontinental wireless telephony possible.

Seeking to promote radio as a new medium, in 1910, De Forest broadcast a live performance by Italian opera star Enrico Caruso from the Metropolitan Opera House. Two years later De Forest found he could boost a weak signal further by feeding the output of one tube to the grid of the next, and so on. He also found that by feeding the output of an Audion tube back to its own grid, he could produce a stable oscillator whose signal could be modulated to carry speech and music.

In the face of a storm of infringement suits, he sold his patents to others to exploit. He went on to invent a system for recording sound on film, making the talkies possible.

REGINALD FESSENDEN (1866 – 1932)

Born in Quebec, Fessenden studied mathematics, but left university without a degree. In 1886, he moved to the US and went to work for Thomas Edison. He worked on a series of projects, but in 1890, when Edison suffered a financial set back, he was laid off. After working in various manufacturing companies, he became professor of electrical engineering at Purdue University, moving onto the Western University of Pennsylvania – now Pittsburgh University – the following year.

From 1900 to 1902, he worked for the Weather Bureau, adapting wireless telegraphy for weather forecasting and storm warnings. In 1900 he was granted a patent for a sensitive detector that made wireless telephone possible and invented the heterodyne receiver which combines two high-frequencies to produce an audible tone. With two Pittsburgh financiers, he formed the National Electric Signaling Company in 1902, which transmitted the first voice signals over a distance. In 1906, he made the first two-way transatlantic transmission. But he fell out with his backers and the company ended up bankrupt.

During his career Fessenden filed some 300 patents, many were subject to litigation. He sued RCA for $60 million, settling out of court in 1928 for a large cash payment. Among his admirers was Elihu Thomson who called Fessenden 'the greatest wireless inventor of the age – greater than Marconi'.

CHAPTER 10

FRESH DREAMS OF FLYING

On my slow return to the normal state of mind, I experienced an exquisitely painful longing after something undefinable. During the day I worked as usual and this feeling, though it persisted, was much less pronounced, but when I retired, the night, with its monstrous amplifications, made the suffering very acute ... my torture was due to a consuming desire to see my mother.

Nikola Tesla[1]

Slowly Tesla recovered. He would go out during the day, as dapper as ever, to have warm compresses applied to his face and his scalp massaged. He looked for new offices and delighted on travelling on the subway which was powered by his induction motors.

In May 1907, Tesla was inducted into the New York Academy of Sciences. He managed to borrow some money and raise a few mortgages, including one with George C. Boldt, the proprietor of the Waldorf-Astoria, where he had not paid rent for 3 years. Just as he seemed to be getting back on his feet came the 'Panic of 1907', when shares plunged 50 per cent from the peak the previous year, causing a run on the banks.

Nevertheless, his imagination was as good as ever. The Wright brothers had made their first powered flight in 1903. Astor was also keen on flying machines and encouraged Tesla to take an interest. However, at a dinner at the Waldorf-Astoria at the beginning of 1908, he made another of his pronouncements:

> The coming year will dispel another error which has greatly retarded the progress of aerial navigation. The aeronaut will soon satisfy himself that an airplane ... is altogether too heavy to soar, and that such a machine, while it has its uses, can never fly as fast as a dirigible balloon. Once this is fully recognized the expert will concentrate his efforts on the latter type, and before many months are passed it will be a familiar object in the sky.[2]

However, he was not entirely wrong as he went on to say: 'Aerial vessels of war will be used to the exclusion of ships.' He also said 'the propeller is doomed'. It would, he said, have to be replaced by 'a reactive jet'.

TESLA'S FLYING MACHINE

Despite his gloomy prognostication, Tesla revealed in *The New York Times* of 8 June 1908 that he was working on a heavier-than-air machine of his own. By 1911, Tesla was ready to spell out his vision in a press interview:

> The flying machine of the future – my flying machine – will be heavier than air, but it will not be an airplane. It will have no wings. It will be substantial, solid, stable. You cannot have a stable airplane. The gyroscope can never be successfully applied to the airplane, for it would give a stability that would result in the machine being torn to pieces by the wind, just as the unprotected airplane on the ground is torn to pieces by a high wind.
>
> My flying machine will have neither wings nor propellers. You might see it on the ground and you would never guess that it was a flying machine. Yet it will be able to move at will through the air in any direction with perfect safety, higher speeds than have yet been reached, regardless of weather and oblivious of holes in the air or downward currents. It will ascend in such currents if desired. It can remain absolutely stationary in the air, even in a wind, for great length of time. Its lifting power will not depend upon any such delicate devices as the bird has to employ, but upon positive mechanical action.[3]

Tesla could not bear being left behind. In 1911, he said: '20 years ago I believed that I would be the first man to fly; that I was on the track of accomplishing what no one else was anywhere near reaching. I was working entirely in electricity then and did not realize that the gasoline engine was approaching a perfection that was going to make the aeroplane feasible.'[4]

His idea, naturally, was to have a plane powered by electricity, with power supplied by stations on the ground. 'I have not accomplished this as yet, but am confident that I will in time,' he said.

THE FLIVVER PLANE TAKES OFF

Tesla applied for a patent on a flying machine in 1921. Dubbed the 'flivver plane' - flivver was early 20th-century slang for a cheap car - it was said to combine the qualities of a helicopter and a plane, and could fly vertically as well as horizontally. According to a press report:

It is a tiny combination plane which, its inventor asserts, will rise and descend vertically and fly horizontally at great speed, much faster than the speed of the planes of today. But despite the feats which he credits to his invention, Tesla says that it will sell for something less than $1,000.

The helicopter-airplane is a small structure, with two wings about 8 ft square. It may have one propeller and it may have several, to be driven by a light but powerful turbine motor of Tesla's invention. When the plane rests on the ground the propeller will be overhead and the wings will be standing vertically. The motor is expected to generate a terrific power that will lift the plane into the air. This power can be sustained only a short while.

At the desired height, the aviator begins to tilt his plane. The wings gradually are brought into a horizontal position that puts the propeller in front of the machine. During this operation the engine power is decreased, and, at this lower power, the helicopter becomes an airplane and is operated as such. The wings now are supporting surfaces, and, except when exceptional speed is wanted for a few moments, the engine will be run at a low rate.

Seats for the pilot and three or four passengers are suspended from trunnions on which they can turn through an angle of 90 degrees. This enables those in the plane to sit in a normal position at all times. With two wheel bases at right angles, the helicopter-airplane is able to descend in a glide or vertically, landing easily either way. As a helicopter, it has a low landing velocity.[5]

Although he obtained more patents for The Flivver Plane in 1928, even by then, he still had not built a full-scale model.

BECOMING YESTERDAY'S NEWS

By 1909, Lee De Forest had perfected the radio-telephone which had been adopted by the navies of Britain, the US and Italy. He had cut out the spark and minimized the chances of the interruption of messages. By transmitting the speaking voice, he claimed, information could now be passed at a rate of 40,000 words an hour instead of 40 words a minute, which was the speed of the fastest Morse operator. Tesla was rapidly becoming a footnote in the history of wireless.

Radio-telephones were being set up on top of tall buildings and, to add insult to injury, radio towers were erected on top of the Waldorf-Astoria. While Tesla took night-time rambles to Grand Central Station which would soon boast its astronomical ceiling, Professor William Pickering announced that he had raised $10,000 to erect a set of mirrors in Texas to send signals to Mars.[6]

ADVANCED BLADELESS TURBINES

Considering his own experiments with wireless transmission 'evidently far in advance of the times', Tesla moved on to other inventions. He came up with a 'bladeless turbine' which, he believed, would replace the petrol engine in a car or could be used to power aircraft, ships or torpedoes. It also worked as a pump.

Just as a rotating magnetic field dragged the rotor around in his AC motors, Tesla believed that it was possible to use steam or compressed air to turn a series of discs attached to a turbine. They utilized the property of viscosity - that is, a fluid's resistance to flow. By adjusting the distance between the discs to match the viscosity and speed of flow of the fluid, Tesla believed that he could create an efficient engine.

In Tesla's engine, fluid entered at the edge of the disc and exited at the shaft in the centre. As the fluid spiralled down between the discs, it dragged them around with it. When the action was reversed, the fluid spiralled out from the centre and acted as a pump or a blower.

Without blades, the engine would be cheaper to build and easier to maintain. What's more it gave a vastly improved power to weight ratio. Tesla claimed that, while the lightest aeroplane engine produces one horsepower (746 watts) for each 2.5 pounds (1.1 kilograms), his engine would produce 25 horsepower (18,642 watts). In his own mind, he had solved the problem of flying.

'I have accomplished what mechanical engineers have been dreaming about ever

WHAT IS VISCOSITY?

All fluids have viscosity. Thick fluids such as molasses have a high viscosity; thin ones, such as air, a low viscosity. All fluids, to a greater or lesser extent, stick to solid surfaces. The molecules near to the surface adhere to it and travel at the same velocity. Molecules a little further away are slowed by a viscous interaction with those stuck on the surface. Further away still, the fluid flows freely. The transition between the layer stuck to the surface and the free-flowing stream is called the boundary layer. Tesla found that he could use 'viscous shear' in the boundary layer to transfer energy from the fluid to the turbine.

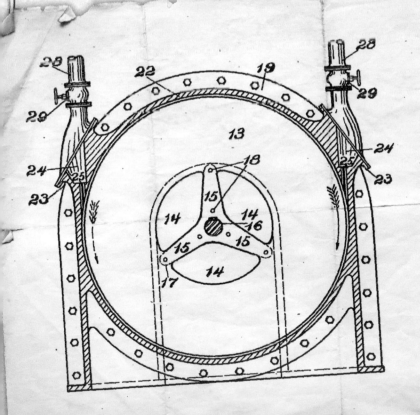

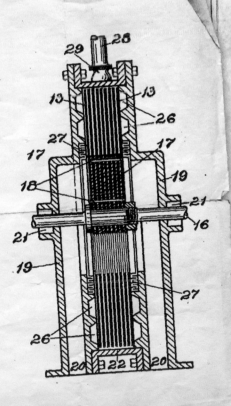

Tesla's original patent diagrams for a turbine engine.

since the invention of steam power,' said Tesla. 'That is the perfect rotary engine.'[7]

That was all very well in theory, but it had taken Tesla years to go from the idea of a rotating magnetic field to the production of a functioning AC motor. Producing a bladeless turbine similarly required a great deal of meticulous engineering. All sorts of different materials and configurations would have to be tested. Nevertheless Tesla was confident that his new engine would be so successful that he would have the money to reopen Wardenclyffe.

COMING UP WITH THE PROTOTYPES

In 1906, Julius Czito, son of Tesla's long-serving assistant Kolman, built the first prototype. It had eight discs each 6 inches (15 cm) in diameter. Weighing less than 10 pounds (4.5 kg) it developed 30 horsepower and would rotate at up to 35,000 revolutions per minute (rpm). At this speed, the metal disks began to distort. Nevertheless, in 1909, Tesla filed two patents - one for the turbine, the other for the pump.

The following year, Czito produced a second prototype with 12 inch (30 cm) discs. With the speed limited to 10,000 rpm, it developed 100 horsepower. A third prototype was slightly smaller. Its discs were 9.75 inches (24.8 cm) in diameter. The speed was limited to 9,000 rpm, but it developed 110 horsepower. Tesla proclaimed that he had invented 'a powerhouse in a hat'.[8] Hooked up to an induction motor it could have made a jet engine.

As he shuttled between various workshops in New York, Providence, Rhode Island, and Bridgeport, Connecticut - where most of the development was done - he became increasingly optimistic.

'I am now at work on new ideas of an automobile, locomotive and lathe in which these inventions of mine are embodied and which cannot help [but] prove a colossal success,' he said. 'The only trouble is to get the cash, but it cannot last very long before my money will come in a torrent.' Later he wrote, undaunted: 'Things are developing very favourably, and it seems my wireless dream will be realized before next summer.'

There is speculation that Tesla installed one of his engines on a 'mysterious craft' that Astor had moored on the Harlem River. 'It seemed to embody an airship with a practical water craft,' said *The New York Times*. However, after a year, it disappeared.[9]

PUTTING ON A PUBLIC SHOW

Astor refused to invest, so Tesla set up the Tesla Propulsion Company with Joseph Hoadley, whose Alabama Consolidated Coal and Iron company planned to use the Tesla pump as a blower in its blast furnaces. Certain his fortunes were on the up again, Tesla took offices in the new Metropolitan Life Tower on Madison Square, then the tallest building in the world.

Eager to promote his new invention, Tesla arranged to give a public demonstration at the Waterside Station of the New York Edison Company. Tesla had two engines built with 18-inch (46-cm) rotors. They were just 3 ft (90 cm) long, 2 ft (60 cm) wide and 2 ft (60 cm) high, and weighed 400 pounds (181 kg). Revolving at 9,000 rpm, they each developed 200 horsepower.

In the demonstration, the two turbines were connected by a torque rod. Turning in the opposite direction, the two motors would engage in a tug-of-war, with the power they developed registered on the torque rod. But in this configuration, with the two turbines pushing against one another, the motors did not actually turn, just strained against each other. The audience was not impressed and the story soon circulated that the test was a complete failure.

BUILDING UP HIGH HOPES

To Tesla, though, the test had been a success. He told the *New York Herald Tribune*: 'One such pump now in operation, with eight discs, 18 inches in diameter, pumps 4,000 gallons a minute to a height of 360 feet.'[10]

As a motor, they would be run using water, air, steam, gas or any other fluid under pressure. 'The motor is especially adapted to automobiles, for it will run on gas explosions as well as on steam,' said Tesla. 'Coupling these engines in series, one can do away with gearing in machinery. Factories can be equipped without shafting.' The applications were limitless.

'With a thousand horse-power engine, weighing only 100 pounds, imagine the possibilities in automobiles, locomotives and steamships,' he said. 'In the space now occupied by the engines of the *Lusitania* 25 times her 80,000 horse power could be developed, were it possible to provide boiler capacity sufficient to furnish the necessary steam.'[11]

The engine would also be perfect to power Tesla's flying machine. Using the current reciprocating petrol engine, he said, 'the aeroplane is fatally defective. It is merely a toy, a sporting plaything. It can never become commercially practical.'

The problem was, of course, he needed money to develop it. Westinghouse was out of business; Astor was not interested. Then in 1912, Astor was lost on the *Titanic*. The following year J.P. Morgan also died. After his funeral, Tesla approached his son Jack for funding. Tesla was still hoping to get Wardenclyffe up and running again, but Jack showed no interest. However, he loaned the inventor $20,000 to develop his turbine.

Tesla then moved into the Woolworth Building, which had taken over from the Metropolitan Life Tower as the world's tallest building. On the wall of his office

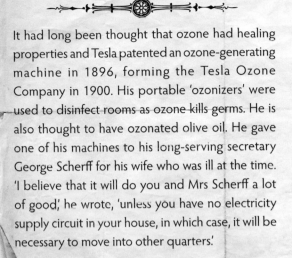

OZONE THERAPY

It had long been thought that ozone had healing properties and Tesla patented an ozone-generating machine in 1896, forming the Tesla Ozone Company in 1900. His portable 'ozonizers' were used to disinfect rooms as ozone kills germs. He is also thought to have ozonated olive oil. He gave one of his machines to his long-serving secretary George Scherff for his wife who was ill at the time. 'I believe that it will do you and Mrs Scherff a lot of good', he wrote, 'unless you have no electricity supply circuit in your house, in which case, it will be necessary to move into other quarters.'

there, Tesla had the famous picture of him in his lab at Colorado Springs quietly reading while huge streaks of artificial lightning crack over his head. The picture is, of course, a cheat, taken using time-lapse photography.

Tesla then tried to sell his turbine to Sigmund Bergmann (1851 – 1927), an old colleague of Edison's who had set up a large manufacturing concern in Germany. But the deal was not concluded before the start of World War I, when Jack Morgan became involved in helping Britain and France to finance the war and lost interest.[12]

MARCONI VERSUS TESLA

Tesla was bitter when Marconi won the Nobel Prize in 1909 for what Tesla considered was his invention. Jack Hammond tried to balance the two men's contributions in his article 'The Future of Wireless' in the *National Press Reporter* in 1912:

Mr Tesla in 1892 showed that the true Hertzian effect was not a means by which it was possible for a sending station to communicate with a receiving station at any great distance. He demonstrated furthermore, that waves propagated at a transmitting station travelled along the ground as a conductor. Today it is acknowledged that these views are correct. It was, however, left to the splendid enterprise of Marconi to crystallize the results of previous investigators into a complete and practical system of space telegraphy ... In 1897, Mr Marconi transmitted messages to a distance of 8.7 miles. Today Mr Marconi says that the maximum effective distance of transmission is 6,000 miles.[13]

REMOTE CONTROL MECHANICAL DOG

A disgruntled Tesla resented that any credit should go to Marconi. He also heard that, while working with Alexander Graham Bell and Tesla's former assistant Fritz Lowenstein, Hammond had invented a mechanical dog that worked by remote control.[14] While Hammond assured Tesla that he had not infringed any of his patents, Tesla insisted that he get a share in any profits. The two of them formed the Tesla-Hammond Wireless Company funded by Hammond's father. Tesla saw this new company as a way to market his bladeless turbine, perhaps to the military. But Hammond was more interested in Tesla's patents on selective tuning, which divided up the radio spectrum into numerous channels - a crucial development in radio.[15]

Hammond contacted the War Department with the proposal for a ship-to-shore communication system and hired Fritz Lowenstein and Benjamin Franklin Miessner to form a military research group at the family estate in Gloucester, Massachusetts.[16] Again Tesla was sidelined.

TESLA'S TALENTED TEAM

FRITZ LOWENSTEIN (1874 - 1922)

Born in Carlsbad in the Austro-Hungarian Empire (now Karlovy Vary in the Czech Republic), Lowenstein studied engineering in Europe before emigrating to the US in 1899 where he went to work for Tesla. He helped build and operate the magnifying transmitter in Colorado Springs. 'Possessed of the highest technical training,' Tesla said, he became a close confidant, discussing the project with him every day over lunch and dinner at the Alta Vista Hotel. They parted when Lowenstein returned to Germany to marry, but Tesla re-employed Lowenstein in 1902 to work at Wardenclyffe. He also worked with Jack Hammond and Alexander Graham Bell, and subsequently began a company making radio sets for the US Navy during World War I, paying royalties to Tesla for the use of his patents.

BENJAMIN FRANKLIN MIESSNER (1890 - 1976)

Miessner studied electrical engineering at Purdue University and worked for the US Navy in 1908 before becoming chief assistant in the Tesla-Hammond lab at Gloucester. He worked on the development of the electric dog and superheterodyne reception. This improved the amplification in a wireless set fifty-fold and allowed them to work without a long aerial, essentially turning the wireless receiver from an experimental apparatus into a domestic appliance. He is also credited with inventing the 'cat's whisker' detector in early crystal sets, which he sold for $200, and the electric organ. A pioneer in aircraft radio and directional microphones in submarines, he sold more than two hundred patents, making over $2 million.

SENDING MESSAGES THROUGH THE AIR

While practical work was underway in Gloucester, Tesla was making more outrageous claims to the convention of the National Electric Light Association in 1911. He said that he would be able to run the streetcars in Dublin by a power station in Long Island City. His wireless transmitter would generate enough power to light the entire United States.

'The current would pass into the air and, spreading in all directions, produced an effect of a strong *aurora borealis*,' he said. 'It would be a soft light, but sufficient to distinguish objects.'[17] *The New York Times* said: 'Queen Isabella of Spain could not have been more amazed when meeting Christopher Columbus to hear about the new world.'[18]

Tesla was even prepared to take on Euclid:

I have annihilated distance with my scheme and when perfected it will not be one mite different than my present plans call for. The air will be my medium, and I will be able to transmit energy of any amount to any place in the world. I will also be able to send messages to all parts of the world, and I will send words out into the world, which will come out of the ground in the Sahara Desert with such force that they can be heard for 15 miles around.

I also hope to set up a central wireless telephone station whereby there will be a force of a million horsepower behind each word uttered into my instrument, and in which distance will play no hindrances whatever. Many hundreds of people will be able to talk at the same time and without any interference with each other.[19]

He also said that he had perfected a new steam engine, a turbine that would produce 10 horsepower but weighed only one pound. The machine would be 'the smallest thing ever seen on wheels and will be more powerful than any automobile engine ever manufactured'.[20]

ENDING THE HAMMOND PARTNERSHIP

Tesla made overtures to the Japanese, hoping that they would take 500 of his turbines to power their torpedoes. He also had meetings with GE and the Seiberling Company, who developed high-speed power boats. Then he worked on prototype car engines, approaching Ford. Kaiser Wilhelm II also took an interest in possible military applications. But Tesla was having problems with the design. The ball bearings wore down too quickly.

As Tesla preferred to work during the night, labour costs soared. Though he did not take a salary himself, in just a few months he found he had laid out $18,000 and asked Hammond for another $10,000 to keep going. But Hammond and Lowenstein were busy installing wireless equipment on naval vessels and competing with De Forest for a $50,000 amplifier deal with AT&T. So Hammond's main interest now was perfecting wireless and he ignored Tesla's request. By then it was clear that it would take a great deal more than $10,000 to perfect the turbine. This effectively ended their partnership.[21]

Hammond spent the money he saved developing remote control, spending around $750,000 dollars on crewless ships, aircraft and submarines. He was eventually compensated by the War Department and did a separate deal with the Radio Corporation of America - later RCA - shortly after Tesla's wireless patents had run out. Yet again the inventor had failed to profit from his inventions. However, they had made Hammond a millionaire in his own right. He used the money to build a faux medieval castle less than a mile from his parents' house. It boasted a nude statue of the famous inventor.[22]

IS IT ELECTRIFIED?

Tesla was never idle for long. An experiment with his Tesla Coils in Sweden had demonstrated that children in an electrified environment grew more quickly and scored higher in aptitude tests. So Tesla went to work for the superintendent of New York's public schools installing Tesla Coils in the walls of a school for a pilot study. The guinea pigs were 50 backward pupils and it was said that 'the brains of the children will receive artificial stimulation to such an extent that they will be transformed from dunces into star pupils'.[23]

For Tesla, there would be limitless applications. *The New York Times* reported:

According to the inventor this experiment in schools will be merely an opening wedge, a suggestion to the people that by the use of high potential electricity they may do away with the use of bromides, phosphates, pepsin tablets, and all kinds of drugs taken for disorders of the nervous and digestive system. If his dream comes true, 10 years from now people will inquire when renting an apartment or a house, 'Is it electrified?' Everyone will have at least one room in his house furnished with a coil generating high-frequency currents. By that time the appliance will be inexpensive, so that at a moderate cost people will be able to obtain superb health and mental brilliancy. Ordinary conversation will then be carried on in scintillating epigrams, and the mental life of the average adult will be so quickened as to equal the brain activity of the most brilliant people living before the time when a generator of high-frequency currents was a household essential.[24]

EXPLAINING STANDING WAVES

A standing wave is caused by the combination of two waves moving in opposite directions and is usually found where a wave is reflected from a surface or the end of a wire. The two waves are superimposed and either add together or cancel each other out. A vibrating rope tied at one end will produce a standing wave. At some positions along the rope there is no movement. These points are called nodes. Either side, where the movement is the greatest, are antinodes.

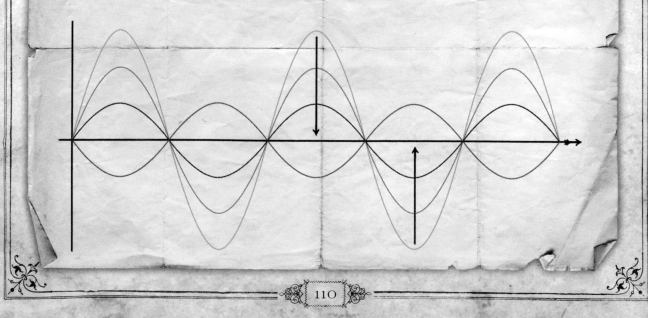

Standing Waves change in amplitude but do not move.

EVERYONE WANTS TO BE TESLA

Tesla also told the newspaper that one of his assistants had been exceedingly stupid, but after a time working around high-voltage equipment the man grew brighter and worked better. Asked about any harmful side-effects, Tesla said that the rays of light issuing from ordinary household incandescent lamps were more harmful than those from a high-voltage coil. He also maintained that the increased prevalence of baldness was due to the effect on the scalp of rays from incandescent bulbs. But the electromagnetic radiation from his coils was, he insisted, perfectly safe.[25]

CONSTANT LEGAL BATTLES

When the *Titanic* sank, Marconi was credited with saving the lives of the 710 survivors, as it was his equipment that had summoned the rescue ships. This was galling for Tesla and other pioneers. So Tesla began suing Marconi for patent infringement. In Britain, Tesla already had let a vital patent lapse. However, Sir Oliver Lodge, the inventor of the coherer, won a suit against Marconi. In France, Tesla succeeded in challenging two of Marconi's patents.[26]

Tesla was sued himself, first by an investor who failed to reap the riches he was promised, then by Westinghouse over equipment they had lent him.[27] Tesla argued that he was not personally liable for this, but agreed to return the equipment.[28] Though his losses were small, he was deeply in debt and the bad publicity damaged his reputation.[29] It seemed he was now fair game. A Mrs Tierstein wanted to shoot him for 'throwing electricity at her'. She was confined to an asylum.[30]

While Telefunken in Germany infringed Tesla's patents, it was too important to sue. But, when the company came to America to

SIR OLIVER LODGE (1851 – 1940)

Lodge entered his father's clay business in Staffordshire, England when he was 14. Then on a visit to London he heard prominent physicist John Tyndall (1820 – 93) lecture at the Royal Institution. This piqued his interest in science and, at the age of 22, he resumed his education. In 1890, the French scientist Édouard Branly (1844 – 1940) showed that iron filings in a glass tube coalesced – or 'cohered' – under the influence of electromagnetic waves. Lodge added a 'trembler' that shook up the filings between waves and made other improvements, making an effective detector.

Following in the footsteps of Hertz, he studied standing waves in conducting wires. After Hertz's untimely death in 1894, he gave a lecture at the Royal Institution called *The Work of Hertz*. When this was published, it had a widespread influence across Europe. He also filed a number of important patents. When his son Raymond was killed in World War I, he became interested in spiritualism and served as president of the Society for Psychical Research.

set up transatlantic stations at Tuckerton, New Jersey, and Sayville, New York, its founder Adolf Slaby (1849 - 1913), sought out Tesla in the hope that they could present a united front to Marconi. However, in 1914, Tesla was also approached by the American Marconi company. But they only offered stock; Tesla needed cash. He appealed to Jack Morgan for help, saying the US government had already installed $10-million-worth of his equipment and he was expecting to receive compensation.[31]

Meanwhile Telefunken was suing Marconi who, in turn, was suing Lowenstein and the US Navy. However, as the wireless equipment Hammond supplied the War Department was being used to test guided missiles, it was classified, so Hammond was immune from litigation.[32]

WORLD WAR I

With the outbreak of World War I in 1914, the British cut Germany's transatlantic cables. Consequently, the Telefunken stations at Tuckerton and Sayville became of vital importance. Fearing that they may be used to direct the movements of battleships and submarines, the British wanted them shut down. While ostensibly neutral, President Woodrow Wilson (1856 - 1924) signed a bill prohibiting radio stations from sending or receiving messages of an 'unneutral nature' and took over the station at Tuckerton.[33]

Although one-tenth of the population of the United States was of German origin, most Americans backed Britain and Tesla's connections with Telefunken made him unpopular. Indeed, he was receiving royalties from their subsidiary, the Atlantic Communication Company, and was giving them advice on how to boost the output of their station at Sayville, which was on Long Island just

a few miles from Wardenclyffe. On 23 April 1915, *The New York Times* reported that its power had been increased from 35 kilowatts to 100 kilowatts and that two 500 ft (150 m) towers were about to be erected, transforming it into one of the most powerful transatlantic communication stations.[34] In a test transmission the previous year, *The New York Times* had received a message from the Burgomaster of Berlin.

TESLA'S DAY IN COURT

Marconi won its case against Lowenstein, but lost its suit against the US Navy. In preparation for the case, the Assistant Secretary for the Navy - later president - Franklin D. Roosevelt (1882 - 1945) reviewed Tesla's 1899 file at the Lighthouse Board. In it was the letter asking Tesla whether he could supply wireless telegraphy apparatus. A review of Marconi's patents was also made.

In 1900, the commissioner of patents John Seymour - who had already upheld Tesla's patents against Michael Pupin's claim that he had invented the AC system - had rejected Marconi's first patent application because of the prior claims of Tesla, Lodge and Ferdinand Braun, who had shared the Nobel Prize for physics with Marconi in 1909. In 1903, the patent office wrote: '... Marconi's pretended ignorance of the nature of a *Tesla oscillator* being little short of absurd... the term *Tesla oscillator* has become a household word on both continents' - that is, Europe and North America.[35]

In 1904, after Seymour had retired, Marconi had been granted a patent. This was being contested in Telefunken's suit against Marconi now going to trial. The *Brooklyn Eagle* reported that some of the world's greatest inventors were on hand to testify[36] - not least Marconi himself. He arrived in New York on the *Lusitania* in

Right: Tesla's Electrical Oscillators published in the *Electrical Experimenter*, July 1919.

Electrical Oscillators

Fig. 2. Small Tesla Coil for gas engine ignition and similar uses

Fig. 3. Tesla Transformer, 12 inch spark, chiefly for wireless

Fig. 5. Later type of Tesla Transformer

Fig. 4. Tesla Oscillator in action, generating undamped waves

Fig. 1. Oscillator with detachable transformer for experimental purposes

Fig. 6. Small oscillator for production of ozone

Fig. 7. Large Tesla Transformer for various purposes

Fig. 8. Tesla Transformer with rotary break for wireless

Fig. 9. Tesla Transformer with mercury interrupter

Fig. 10. Large Tesla Transformer with hermetically sealed mercury interrupter

Fig. 12. Another type of Tesla Transformer with sealed mercury interrupter

Fig. 11. Tesla Transformer with sealed mercury interrupter for low tension work

April 1915, telling reporters that he had seen the periscope of a German submarine on the crossing.[37] The press were on his side as Italy sided with the Allies in World War I.

Also appearing for the defence was Tesla's old adversary, Columbia Professor Michael Pupin, who shocked the court by saying: 'I invented wireless before Marconi or Tesla and it was I who gave it unreservedly to those who followed!'[38]

A local newspaper reported that Tesla was so shocked that 'watching his fellow Serb upon the stand, Tesla's jaw dropped so hard, it almost cracked upon the floor'.[39]

Pupin had already made it clear where he stood. In the press he said of Marconi: 'His genius gave the idea to the world, and he taught the world how to build a telegraphic practice upon the basis of this idea ... In my opinion, the first claim for wireless telegraphy belongs to Mr Marconi absolutely, and to nobody else.'[40]

TESLA THE TRAILBLAZER

While Pupin could only assert that Marconi was the inventor, Tesla came armed with lectures, articles and patents. While Marconi had only been granted his first US patent in 1904, Tesla had given a practical demonstration of wireless in St Louis, Missouri in 1893. He had transmitted a signal from his Houston Street laboratory to West Point before 1897. People who visited his lab saw the equipment. He then compared the Marconi patent to his own, saying: 'If you take these two contemporaneous diagrams, and examine the subsequent developments, you will find that absolutely not a vestige of that apparatus of Marconi remains, and that in all the present system there is nothing but my four-tuned circuits. Everybody is using them.'[41]

Following Tesla to the stand was John Stone Stone, who could himself claim to be the inventor of the radio. He filed a crucial patent on tuning 2 years before Marconi, and he acknowledged that it was Tesla, not Marconi, who was a trailblazer.

Due to World War I, the legal battle was abandoned before the case was decided. Fearing a German attack, Marconi did not return to Europe on the *Lusitania*. Instead under a false name and in heavy disguise, he set sail on the *St Paul*. On 7 May 1915, the *Lusitania* was sunk by a German submarine with the loss of 1,134 lives, including 128 US citizens. America was outraged. It was already suspected that the Telefunken radio station was passing coded messages to submarines. Two weeks later *The New York Times* reported that the Germans had developed 'air torpedoes' which were dropped from zeppelins and controlled by wireless[42] - the very thing Tesla had claimed to have invented.

When the US joined World War I in 1917, the former head of the Tuckerton station was arrested for spying and the Sayville station was taken over by the government. Tesla's monthly royalty cheques from Telefunken stopped. Edison, Fessenden, Pupin, Thomson and others got jobs as advisors to the government, but Tesla - possibly because of the connections with Telefunken - was left out.[43]

Tesla demonstrates wireless power transmission in his Houston Street laboratory, 1899.

MICHAEL PUPIN (1858 – 1935)

Born in Banat, a buffer zone between the Ottoman and the Austrian Empires, Pupin was a Serb like Tesla. His parents were illiterate and sent him to Prague. After a year, not yet 16, he went alone to America, arriving in New York in 1874. For 5 years, he took a series of odd jobs, while studying at night for admission to Columbia College, now Columbia University. He went on to study in Cambridge and Berlin, where he worked under Helmholtz. He returned to New York to teach mathematical physics at the newly formed department of electrical engineering. In 1901, he was made Professor of Electromechanics, a position he held until he retired in 1931.

In 1896, he discovered that atoms struck by X-rays emit secondary X-ray radiation and worked on X-ray fluoroscopy. Five years later, the Bell Telephone Company bought the patents for his method of extending the range of telephone communication by placing loading coils at specific distances along the line. In 1924, he was awarded a Pulitzer Prize for his best-selling autobiography *From Immigrant to Inventor*.

Pupin sided with Elihu Thomson in the controversy over who invented the AC polyphase system and Tesla accused Pupin of stealing his work. In the long passages on the development of AC in *From Immigrant to Inventor*, Tesla is hardly mentioned.

When working with X-rays, Pupin again ignored Tesla's contribution. It was Pupin who introduced Marconi to Tesla in 1900, but he also helped facilitate Marconi's cooperation with Edison, earning him, once more, the enmity of Tesla.

When Pupin was on his deathbed in 1935, he got his secretary to visit Yugoslav diplomat Stanko Stoilkovic and ask him to plead with Tesla to visit Pupin who wanted to make peace with him before he died. Tesla said that he would have to think about Pupin's request overnight. The following day, Tesla turned up at the hospital. In Pupin's room he approached the bed with his hand extended and said: 'How are you old friend?'

Pupin was overcome with emotion. They were left alone to talk. Tesla said they had agreed they would meet up again, but Pupin died immediately after Tesla's visit. Reconciled at last, Tesla attended his funeral.

KARL FERDINAND BRAUN (1850 – 1918)

Born in Germany, Braun received his doctorate from the University of Berlin in 1872 and held a number of academic posts before becoming director of the Physical Institute and Professor of Physics at the University of Strasbourg in 1895. In 1897, he developed the first oscilloscope, or Braun tube, to study alternating currents using a beam of electrons in a cathode ray. From this, television tubes were developed. He went on to study why early wireless transmission was limited to 9.5 miles (15 km), concluding that the limiting factor was the length of the spark. The solution was to introduce a sparkless antenna circuit, which he patented in 1899. He also developed an antenna that directed the transmission in one direction. The Nobel Committee recognized that he had made considerable improvements to Marconi's apparatus and awarded the Nobel Prize to them jointly in 1909.

Braun travelled to New York in 1915. When the US joined the Allies in World War I in 1917, he was detained as an enemy alien and died before the war ended.

JOHN STONE STONE (1869 – 1943)

Born in America, but brought up in Egypt and Europe, Stone, who was fluent in Arabic, French and English, was brought home to the US to study in the school of mines at Columbia and Johns Hopkins University, before entering the Bell Labs in Boston. In 1899, he set up the Stone Telegraph and Telephone Company. Lecturing on electrical oscillations at the Massachusetts Institute of Technology, he filed a patent on tuning in 1902. He also developed a wireless direction finder, worked on the use of loading coils on telephone lines before Pupin and became president of the Institute of Radio Engineers and the AIEE. He holds many 'space telegraphy' patents.

Left: Michael Pupin.

Above: Karl Ferdinand Braun.

CHAPTER 11

THE NOBEL PRIZE

I have concluded that the honour has been conferred upon me in acknowledgement of a discovery announced a short time ago which concerns the transmission of electrical energy without wires. This discovery means that electrical effects of unlimited intensity and power can be produced, so that not only can energy be transmitted for all practical purposes to any terrestrial distance, but even effects of cosmic magnitude may be created.

Nikola Tesla[1]

On 6 November 1915, *The New York Times* announced that Tesla and Edison had been awarded the Nobel Prize for physics. The source of the story, apparently, was the Copenhagen correspondent of the *Daily Telegraph*. Tesla had received no notification, but when interviewed by the *Times* he said that he thought he had been given the Nobel Prize for a device he had filed a patent for a month earlier that made it 'practicable to project the human voice not only for a distance of 5,000 miles, but clear across the globe. I demonstrated this in Colorado in 1899.'[2] He went on to explain how it would work:

The plant would simply be connected with the telephone exchange of New York City and a subscriber will be able to talk to any other telephone subscriber in the world, and all this without any change in his apparatus. This plan has been called my 'world system'. By the same means I propose also to transmit pictures and project images, so that the subscriber will not only hear the voice, but see the person to whom he is talking ...

A further advantage would be that the transmission is instant and free of the unavoidable delay experienced with the use of wire and cables. As I have already made known, the current passes through the earth, starting from the transmission station with infinite speed, slowing down to the speed of light at a distance of 6,000 miles, then increasing in speed from that region and reaching the receiving station again with infinite velocity.

It's all a wonderful thing. Wireless is coming to mankind in its full meaning like a hurricane some of these days. Some day there will be, say, six great wireless telephone stations in a 'world system' connecting all the inhabitants of this earth to one another not only by voice but by sight. It's surely coming.[3]

ILLUMINATING THE SKY

This discovery had a direct bearing on the problems uppermost in the public's mind, he said, the perfection of wireless telephony. On 7 November, he told the *Times*:

We will deprive the ocean of its terrors by illuminating the sky, thus avoiding collisions at sea and other disasters caused by darkness. We will draw unlimited quantities of water from the oceans and irrigate the deserts and other arid regions. In this way we will fertilize the soil and derive any amount of power from the sun. I also believe that ultimately all battles, if they should come, will be waged by electrical waves instead of explosives.

He would say nothing further on the matter. However, he agreed that Edison deserved a dozen Nobel Prizes, though he said he had no idea which one he had been awarded the prize for. When shown the despatch, Edison wisely declined to comment. Meanwhile the idea of winning the prize had gone to Tesla's head. He wrote to the Johnsons:

In a thousand years, there will be many recipients of the Nobel Prize. But I have not less than four dozen of my creations identified with my name in the technical literature. These are honours real and permanent, which are bestowed not by a few who are apt to err, but by the whole world which seldom makes a mistake, and for any of these I would give all the Nobel prizes during the next thousand years ... Josie will never had the chance of turning me away as a beggar, but I shall give her soon, the opportunity of slamming your door in the face of a millionaire.[4]

This was no joke to Robert Johnson who had fallen on hard times. He wrote later: 'When that Nobel Prize comes, remember that I am holding onto my house by the skin of my teeth and desperately in need of cash!'[5]

As it was, neither Tesla nor Edison was awarded the prize. That year the Nobel

Prize for physics went to William Bragg and his son Lawrence for their research into crystalline structures using X-rays. Tesla had not even been nominated, though Edison was. Tesla was not nominated until 1937 and did not get it then either.

However, Tesla's long-standing friend and first biographer John O'Neill said that Tesla actually turned down the award. 'To have the award go first to Marconi, and then to be asked to share the award with Edison, was too great a derogation of the relative value of his work to the world for Tesla to bear without rebelling,'[6] he wrote in *Prodigal Genius*. According to O'Neill, Tesla did not put himself in the same category as Edison. He considered himself a discoverer of new principles, while Edison was an inventor who exploited new discoveries for commercial gain.[7]

THE BOLTS OF THOR

Although *The New York Times* had announced that the Braggs had won the Nobel Prize on 14 November 1915, it continued to say that Tesla was a 1915 Nobel Prize winner in an article on 8 December, headlined *Tesla's New Device Like Bolts of Thor*, when the paper reported that he was taking out a patent on a 'manless airship'. It had neither an engine nor wings and could be sent at a speed of 300 miles (480 km) a second to any place on the globe using electricity. According to the *Times*:

> Ten miles or a thousand miles, it will all be the same to the machine, the inventor says. Straight to the point, on land or on sea, it will be able to go with precision, delivering a blow that will paralyze or kill, as is desired. A man in a tower on Long Island could shield New York against ships or army by working a lever ...[8]

Tesla refused to go into further details. However, he dismissed electrical engineer Charles H. Harris's suggestion that, in time of war, the country would be surrounded by 'an electrical wall of fire' as 'not practical' as it would take more than all the generators in the US to power it.[9]

DESCRIBING RADAR

While Tesla's ideas on unmanned airships and bolts of Thor seem unworldly, he also described a way of detecting ships at sea. His idea was to transmit high-frequency radio waves that would reflect off the hulls of vessels and appear on a fluorescent screen. In 1917, he said: 'We may produce at will, from a sending station, an electrical effect in any particular region of the globe; we may determine the relative position or course of a moving object, such as a vessel at sea, the distance traversed by the same, or its speed.'[10] This was one of the first descriptions of what we now call radar. Again it was too far ahead of its time to be taken seriously. However, in 1934 the French engineer Émile Girardeau (1882 - 1970) built an obstacle-locating radio apparatus - 'conceived according to the principles stated by Tesla,'[11] he said - and obtained a patent for a working system, part of which was installed on the liner *Normandie* in 1935.

THE EDISON MEDAL

In 1917, Tesla was awarded the Edison Medal by the American Institute of Electrical Engineers. According to O'Neill, Tesla was reluctant to accept it at first, but was persuaded to do so by the chairman of the medal committee, Bernard A. Behrend, who was an admirer and close personal friend.[12] Tesla turned the medal down initially because it was nearly 30 years since he had announced his rotating electric field and his AC system to the Institute.[13] 'I do not

THE EDISON MEDAL

THOMAS A·EDISON *and the* EDISON MEDALISTS

The Edison Medal was created in 1904 by a group of Edison's friends and associates as an annual award to be given to a living electrician for 'meritorious achievement in electrical science and art'. In 1909, the American Institute of Electrical Engineers agreed to present it as their highest award. The first recipient was Tesla's rival Elihu Thomson. George Westinghouse and Alexander Graham Bell also received the award. The medal is now presented by the Institute of Electrical and Electronics Engineers, formed when the AIEE merged with the Institute of Radio Engineers.

The Edison Medal recipients 1909 - 1922.

need its honours and someone else may find it useful,'[14] he said. Pressed by Behrend for further explanation, Tesla said:

> *You propose to honour me with a medal which I could pin on my coat and strut for a vain hour before the members and guests of your Institute. You would bestow an outward semblance of honouring me but you would decorate my body and continue to let it starve, for failure to supply recognition, my mind and its creative products which have supplied the foundation upon which the major portion of your Institute exists. And when you would go through the vacuous pantomime of honouring Tesla you would not be honouring Tesla but Edison who had previously shared unearned glory from every previous recipient of this medal.*[15]

Despite his rancour, Tesla was cajoled into accepting the Medal. After all it could hardly be awarded to Edison. But the acceptance speech presented Tesla with something of a problem. When he had addressed the AIEE in 1888, he had a lab where he could prepare his demonstrations. Now he had none. Nor could he expect to equal the lectures he had taken on the road in the 1890s. He had no props.

After a private dinner at the Engineers' Club, the medal winner was to give a formal address in the auditorium of the United Engineering Societies Building. However, as the members of the Institute assembled there, Tesla was nowhere to be seen. Behrend found him feeding the pigeons in the plaza of New York Public Library. As Behrend approached, Tesla had pigeons perched on his head, shoulders and arms, and he had a carpet of them pecking at seed around his feet. It was clear that the pigeons meant more to him than the members of AIEE. Behrend begged Tesla not to let him down.[16]

THE WHEELS OF INDUSTRY WILL CEASE

In the auditorium of the United Engineering Societies Building, Behrend said that Tesla had been taken temporarily unwell, but he was now okay and the proceedings would be delayed by about 20 minutes. When the presentation began Dr Arthur Kennelly from the Edison company said that Tesla was being awarded the Edison Medal for the development of rotating magnetic fields, which had made it possible to use AC in electric motors, and for his investigations into high-frequency currents.

Charles A. Terry, who had worked with Tesla on some of his early research, ran through Tesla's achievements to date. Behrend followed up by pointing out that, by an extraordinary coincidence, Tesla had given the first lecture on polyphase AC there exactly 29 years earlier, adding:

> *Not since the appearance of Faraday's experimental researches in electricity has a great experimental truth been voiced so simply and so clearly as this description of Mr Tesla's great discovery of the generation and utilization of polyphase alternating currents. He left nothing to be done for those who followed him. His paper contained the skeleton even of the mathematical theory.*
>
> *Three years later, in 1891, there was given the first great demonstration, by Swiss engineers, of the transmission of power at 30,000 volts from Lauffen to Frankfurt by means of Mr Tesla's system. A few years later this was followed by the development of the Cataract Construction Company, under the presidency of our member, Edward D. Adams, and with the aid of the engineers of the Westinghouse Company. It is interesting to recall here tonight that in Lord Kelvin's report to Mr Adams, Lord Kelvin recommended the use of direct current for the development*

of power at Niagara Falls and for its transmission to Buffalo.

The due appreciation or even enumeration of the results of Mr Tesla's invention is neither practicable nor desirable at this moment. There is a time for all things. Suffice it to say that, were we to seize and to eliminate from our industrial world the results of Mr Tesla's work, the wheels of industry would cease to turn, our electric cars and trains would stop, our towns would be dark, our mills would be dead and idle. Yea, so far-reaching is this work that it has become the warp and woof of industry ... His name marks an epoch in the advance of electrical science. From that work has sprung a revolution in the electrical art.[17]

Behrend then asked Tesla to accept the Medal, not for the purposes of perpetuating his name - 'the name of Tesla runs no more risk of oblivion than does that of Faraday, or that of Edison'. Nor was the Medal evidence that Tesla's work had received official sanction - 'his work stands in no need of such sanction'.[18]

No, Mr Tesla, we beg you to cherish this medal as a symbol of our gratitude for the new creative thought, the powerful impetus, akin to revolution, which you have given to our art and to our science. You have lived to see the work of your genius established. What shall a man desire more than this? There rings out to us a paraphrase of Pope's lines on Newton: 'Nature and Nature's laws lay hid in night. God said, Let Tesla be, and all was light.'[19]

TESLA'S ACCEPTANCE SPEECH

Accepting the Edison Medal, Tesla said he was grateful for the sympathy and appreciation shown him. Great strides had been made in the transmission and transformation of energy, he said, but 'we are pressing on, inspired with the

BERNARD A. BEHREND (1875 – 1932)

Born in Villeneuve, Switzerland, Behrend studied engineering in Berlin before emigrating to the US in 1894. He was one of the first to understand Tesla's work with alternating currents and began publishing articles on the applications of AC in 1896. His treatise *The Induction Motor* was published by *Electrical World & Engineer* in 1901. It was expanded and published as a book, *The Induction Motor and Other Alternating Current Motors* in 1921. Behrend met Tesla in 1901 when he was assigned to design a motor for Wardenclyffe. In Behrend, Tesla found a like-mind. Behrend was the inventor of numerous electrical devices and took out over 70 patents. After a period of ill-health he committed suicide at his home in Massachusetts.

hope and conviction that this is just the beginning, a forerunner of further and still greater accomplishments'.[20] He had not written an acceptance speech and spoke off the cuff, saying:

I come from a very wiry and long-lived race. Some of my ancestors have been centenarians, and one of them lived 129 years. I am determined to keep up the record, and believe there is a prospect of accomplishing it. Then, nature has given me a vivid imagination which, through incessant exercise and training, through the study of scientific subjects, and the verification of theories through experiment, has become very accurate in results, so that I have been able to dispense, to a large extent, with the slow labours, wasteful and expensive processes of practical development of the ideas I conceive. It has made it possible for me to

explore extended fields with great rapidity and get results with the least expenditure of vital energy. By this means, I may tell you also, I am able to picture the objects of my desires in forms so real and tangible that I can rid myself of that morbid craving for perishable possessions to which so many succumb.

My life was also wonderful in another respect, for physical endurance or energy. If you inquire into the career of successful men in the inventor's profession, you will find, as a rule, that they are as remarkable for their physical as for their mental capacities ... When I turned my thoughts to inventions, I found that I could visualize my conceptions with the greatest facility. I did not need any models and drawings or experiments, I could do it all in my mind, and I did. The way I unconsciously evolved what I considered a new method in materializing inventive concepts and ideas, is exactly opposed to the purely experimental method, of which undoubtedly Edison is the greatest and most successful exponent.

The moment you construct a device to carry into practice a crude idea you will find that you will be engrossed with the details and effects of the apparatus. As you go on changing and constructing, you will lose the forces of concentration, and you will lose sight of the great underlying principle. You obtain results, but at the sacrifice of quality. I did not construct. When I got an idea, I started right away to build it up in my mind. I changed the structure, I made improvements, I experimented, and I ran the device in my mind.

It is absolutely the same to me whether I place my turbine in my mind or have it in my shop actually running in my test. It makes no difference. The results are the same. In this way you see I can develop and perfect an invention without touching anything, and when I have gone so far that I have put into that device every possible improvement I can think of, that I can see no fault in it any more, I then construct it, and every time my device works as I conceived it would, my experiment comes out exactly as I plan it, and in 20 years there has not been a single, solitary experiment which did not come out exactly as I thought it would.[21]

ARTHUR E. KENNELLY (1861 – 1939)

Born in India of Irish parents, Kennelly studied in London before going to work for Edison in New Jersey as a mathematician. With Harold P. Brown, he worked on the development of the electric chair. He also developed the use of complex numbers in analyzing AC circuits. In 1901, he noticed that Marconi's signals arrived in Newfoundland in greater strength than expected and postulated that they had been reflected from an ionized layer in the upper reaches of the atmosphere predicted by English electrical-engineer Oliver Heaviside (1850 – 1925). This became known as the Kennelly-Heaviside layer. Professor of Electrical Engineering at Harvard, Kennelly served as president of the AIEE (1898 – 1900) and the Institute of Radio Engineers (1916). He was awarded the Edison Medal in 1933.

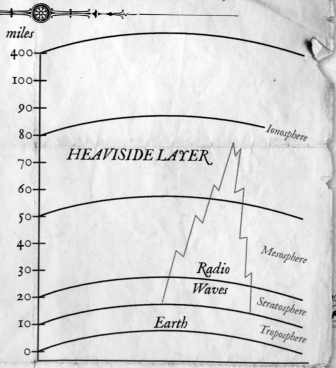

CHAPTER 12

DESTROYING THE DREAM

I have received reports which have completely confounded me all the more as I am now doing important work for the government with a view of putting the plant to a special moment... I trust that you will appreciate the seriousness of the situation and will see that the property is taken good care of and that all apparatus is carefully preserved.

Nikola Tesla[1]

Tesla had signed over Wardenclyffe to the Waldorf-Astoria as he could not pay his hotel bill which had now reached $20,000 ($400,000 at today's prices). However, he still hoped it would be returned when he raised the money to pay the bill, but the hotel management was determined to demolish the tower and sell off parts. Tesla's response was to emphasize the tower's usefulness in the event of war. It was, after all, home to his death ray and could shatter armies with its 'Bolts of Thor'.[2] Nevertheless, he was told that the demolition of the tower was to go ahead. With the US now in World War I, there would be no more money. As part of the war effort, Westinghouse, American Marconi and AT&T were allowed to pool their patents and produced each other's equipment without compensating the original inventors. 'A great wrong has been done,' he wrote later, 'but I am confident that justice will prevail.'[3]

In July 1917, Tesla left the Waldorf-Astoria where he had lived for 20 years. After persuading the management to let him keep many of his personal effects in the basement, he took the train to Chicago where he planned to continue work on his bladeless turbines. There he moved into the Blackstone Hotel next door to the University of Chicago.[4]

The following month, Tesla received a letter from George Scherff, his secretary, telling him that explosives experts had placed charges on major struts of the tower at Wardenclyffe and had blown them up.[5]

SUSPICIONS OF ESPIONAGE

Meanwhile the story was circulated that suspected German spies had been using the tower for radio communication. The *Electrical Experimenter* said: 'Suspecting that German spies were using the big wireless tower erected at Shoreham, L.I., about 20 years ago by Nikola Tesla, the Federal Government ordered the tower destroyed and it was recently demolished with dynamite. During the past month several strangers had been seen lurking about the place.'[6]

And the *New York Sun* gleefully reported: 'The destruction of Nikola Tesla's famous tower ... shows forcibly the great precautions being taken at this time to prevent any news of military importance getting to the enemy.'[7]

Tesla was upset by the implication that he was disloyal to his new country. He had argued that the structure should have been preserved to help locate and destroy enemy submarines. If the tower had been destroyed to curb spying, Tesla pointed out that he should have been compensated by the government for the large amount of money he had put into it. As it was, he made no public protest when the US was at war. However, 2 years later he wrote that his dream had been destroyed by rivals, saying:

I am unwilling to accord to some small-minded and jealous individuals the satisfaction of having thwarted my efforts. These men are to me nothing more than microbes of a nasty disease. My project was retarded by the laws of nature. The world was not prepared for it. It was too far ahead of time, but the same laws will prevail in the end and make it a triumphal success.[8]

TELEPHONY TAKES OVER

With Tesla's World Telegraphy Centre now in pieces, representatives of American Marconi, AT&T, Westinghouse and GE got together behind closed doors in Washington and formed RCA. At the end of the war, radio stations were returned to their rightful owners, favouring RCA.

Using Marconi patents, Westinghouse set up independently. In 1920, Tesla wrote, offering his services. They were refused. However, a little later, Westinghouse wrote again, asking Tesla if he would like to broadcast to their 'invisible audience' one Thursday evening.[9]

Tesla replied that, 20 years earlier, he had promised his friend J.P. Morgan that his 'world system' would enable the voice of a telephone subscriber to be transmitted to any point on the globe. 'I prefer to wait until my project is completed before addressing an invisible audience,'[10] he said proudly.

THE SCIENCE FANTASY FACTOR

Though Tesla's tower was in ruins, the idea would live on. Before leaving New York, Tesla teamed up with long-term admirer Hugo Gernsback, the editor of *Electrical Experimenter*. He had met Tesla in 1908 when he visited his lab to see a bladeless turbine. Eleven years later, Gernsback recorded his impressions in *Electrical Experimenter*:

The door opens and out steps a tall figure – over 6 ft high – gaunt but erect. It approaches slowly, stately. You become conscious at once that you are face-to-face with a personality of a high order. Nikola Tesla advances and shakes your hand with a powerful grip, surprising for a man over 60. A winning smile from piercing light blue-grey eyes, set in extraordinarily deep sockets, fascinates you and makes you feel at once at home.

You are guided into an office immaculate in its orderliness. Not a speck of dust is to be seen. No papers litter the desk, everything just so. It reflects the man himself, immaculate in attire, orderly and precise in his every movement. Dressed in a dark frock coat, he is entirely devoid of all jewellery. No ring, stickpin, or even watch-chain can be seen.

Tesla speaks – a very high almost falsetto voice. He speaks quickly and very convincingly. It is the man's voice chiefly which fascinates you.

As he speaks you find it difficult to take your eyes off his own. Only when he speaks to others do you have a chance to study his head, predominant of which is a very high forehead with a bulge between the eyes – the never-failing sign of an exceptional intelligence. Then the long, well-shaped nose, proclaiming the scientist.

How does this man, who has accomplished such tremendous work, keep young and manage to surprise the world with more and more new inventions as he grows older?

How does this youth of sixty, who is a professor of mathematics, a great mechanical and electrical engineer and the greatest inventor of all times, keep his physical as well as remarkable mental freshness? [11]

Gernsback employed the artist Frank R. Paul to show the world what Tesla's tower would have looked liked if it had been completed. For the cover of the *Electrical Experimenter* Paul added transmitters and Tesla's wingless flying machines zapping nearby ships with their death-rays. Tesla was so thrilled, he used the illustration as his letterhead.

In 1919, *Electrical Experimenter* serialized Tesla's autobiography *My Inventions*. This too was illustrated by Frank Paul's drawing, along with photographs of the equipment. This boosted the circulation of the magazine to around 100,000 and provided Tesla a modest income. However, Tesla felt he had been underpaid and when Gernsback sought to put him on the cover of *Electrical Experimenter* again, Tesla refused, saying: 'I appreciate your unusual intelligence and enterprise, but the trouble with you seems to be that you are thinking only of H. Gernsback first of all, once more, and then again.'[12]

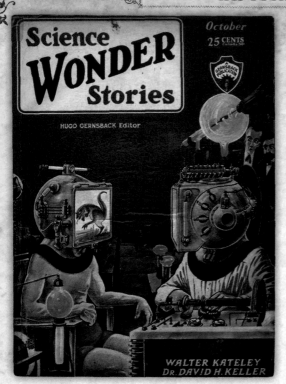

Nevertheless, Gernsback never stinted in his praise of Tesla, running his articles in several of his magazines. He even latched on to some of Tesla's more outlandish ideas. While Tesla did not believe in extra-sensory perception or mind-reading in the psychic sense, he did think it possible to read another person's thoughts by attaching television equipment to their retina. Paul depicted this on the cover of *Science Wonder Stories* in October 1929 which showed two people wearing thought-reading helmets.

THE CHICAGO YEARS

Tesla stayed in Chicago until the end of World War I. He worked on his turbines at Pyle National. Again he refused a salary, hoping to make bigger profits from his inventions in the end. But earning nothing from his wireless patents, his only source of income was the Waltham Watch Company who were manufacturing a speedometer he had designed.

THE BIG SHIP

Tesla was still depending on his turbines to make him rich 'within four months', he told Jack Morgan before he left New York. 'My big ship is still to come in,' he said, 'but I have a marvellous opportunity having perfected an invention that will astound the whole world.'[13]

But he had yet to perfect it. The high rotational speed put too much stress on the discs which risked cracking. Various alloys were tried. However, the advantages of his turbine was obvious. 'Suppose that the steam pressure of the locomotive would vary from say 50 to 200 pounds, no matter how rapidly,' he wrote, 'this would not have the slightest effect on the performance of the turbine.'[14]

The US Machine Manufacturing Company asked about putting one in an aeroplane. The Chicago Pneumatic Tool Company also made enquiries. He told George Scherff that he was expecting to make $25 million a year from his turbines. All the while the debts kept piling up.[15]

MADE IN MILWAUKEE

Tesla finally succeeded in getting the Allis-Chalmers Manufacturing Company of Milwaukee interested. They made reciprocating engines, turbines and other heavy machinery. However, he displayed a lack of tactic and diplomacy that ruined the project from the outset. Insisting on entering negotiations with the senior staff, he went directly to the president of the company and, while his engineers were preparing a feasibility report, Tesla contacted the board of directors and sold them the idea before the engineers had had their say.

Three of Tesla's turbines were built. Two had 20 discs that were 18 inches (46 cm) in diameter. Tested with 80 pounds

Above: Frank R. Paul's illustration of Tesla's thought-reading helmets on the cover of *Science Wonder Stories*, October 1929.

THE SCI-FI CONNECTION

HUGO GERNSBACK (1884 – 1967)

Born Hugo Gernsbacher in Luxembourg, Gernsback had heard of Tesla as a child. He studied electronics in Bingen Technicum in Germany, before emigrating to the US in 1903. He imported electronic components from Europe and, in 1908, founded the magazine *Modern Electrics*. The *Electrical Experimenter* followed in 1913 and *Science and Invention* in 1920. These magazines began to carry science-fiction stories, starting with his own *RALPH 124C41+* set in the year 2660, which was serialized in *Modern Electronics* in 1911.

He began the first dedicated science-fiction magazine, *Amazing Stories*, in 1926. Despite his reputation for dubious business practices, he continued to write and publish. The Hugo Awards, presented annually by the World Science Fiction Convention, were named after him and, in 1960, he received a special Hugo Award as the 'Father of Magazine Science Fiction'.

FRANK R. PAUL (1884 – 1963)

One of the most influential science fiction illustrators of his time, Frank Paul was born in Austria and studied art in Vienna, Paris and New York. He was working as an illustrator on a rural newspaper when Gernsback employed him to work on *Electrical Experimenter*. He produced the cover illustration for Gernsback's *RALPH 124C41+* when it appeared in book form in 1925. He also worked for *Amazing Stories*, *Science Wonder Stories*, *Planet Stories*, *Superworld Comics*, *Science Fiction* magazine and *Marvel Comics*. Paul is credited with the first depiction of a flying saucer, a space ship and a space station.

Frank R. Paul's science fiction illustration of the Wardenclyffe Tower.

pressure, they developed 200 horsepower at between 10,000 and 12,000 rpm. This had the same output as Tesla's 1911 model with discs half the diameter when operating at 125 pounds pressure and 9,000 rpm. They also built a larger version.[16] Hans Dahlstrand, consulting engineer of the steam turbine department, wrote a report saying:

> We also built a 500 kilowatt steam turbine to operate at 3,600 revolutions. The turbine rotor consisted of 15 disks 60 inches [152 cm] in diameter and one-eighth of an inch thick [3 mm]. The discs were placed approximately one-eighth inch apart. The unit was tested by connecting to a generator. The maximum mechanical efficiency obtained on this unit was approximately 38 per cent when operating at steam pressure of approximately 80 pounds absolute and a back pressure of approximately 3 pounds absolute and 100°F [38°C] superheat at the inlet. When the steam pressure was increased above that given the mechanical efficiency dropped, consequently the design of these turbines was of such a nature that in order to obtain maximum efficiency at high pressure, it would have been necessary to have more than one turbine in series.[17]

Dahlstrand reported that difficulties were encountered in the Tesla turbine from vibration, making it necessary to re-enforce the discs, and that this difficulty is common to all turbines. Vibration cracked wheels and wrecked turbines, sometimes within a few hours and sometimes after years of operation. This vibration was caused by taking such terrific amounts of power from relatively light machinery. The Dahlstrand Report identified other problems:

> The efficiency of the small turbine units compares with the efficiency obtainable on small impulse turbines running at speeds where they can be directly connected to pumps and other machinery. It is obvious, therefore, that the small unit in order to obtain the same

efficiency had to operate at from 10,000 to 12,000 revolutions and it would have been necessary to provide reduction gears between the steam turbine and the driven unit.

> Furthermore, the design of the Tesla turbine could not compete as far as manufacturing costs with the smaller type of impulse units. It is also questionable whether the rotor disks, because of light construction and high stress, would have lasted any length of time if operating continuously.

> The above remarks apply equally to the large turbine running at 3,600 revolutions. It was found when this unit was dismantled that the discs had distorted to a great extent and the opinion was that these discs would ultimately have failed if the unit had been operated for any length of time.

> The gas turbine was never constructed for the reason that the company was unable to obtain sufficient engineering information from Mr Tesla indicating even an approximate design that he had in mind.[18]

TERMINATING THE TURBINE

Tesla seems to have walked out at this stage. Later Tesla was asked why he stopped working with Allis-Chalmers. He said: 'They would not build the turbines as I wished.' But he would not elaborate. Allis-Chalmers went on to manufacture a different type of gas turbine that was in production for years.[19]

A number of engineers had tried to explain the failure of Tesla's turbine. One expert said that, while being a fine concept and an excellent machine, it was not that much better than other designs. Another said that not enough money had been spent on research and development. Metallurgy was in its infancy and the instrumentation for measuring its performance had not been developed, nor had the magnetic bearing it would have needed to run efficiently.[20] However, manufacturers have made pumps using Tesla's principles and others have experimented with making the discs using advanced materials such as carbon fibre, titanium-impregnated plastic and Kevlar.

In disappointment, Tesla returned to Colorado Springs where he conducted some experiments in the lab of the local engineering school and relived old times.

After a sojourn at Waltham Watches in Boston, he worked on a petrol-powered turbine at Budd Manufacturing in Philadelphia. And he was not without his successes. He sold a motor that was used in cinema equipment to Wisconsin Electric and a 'fluid diode' to an oil company that was said to be 'the only valving patent without moving parts'.[21] Money began coming in, but never in the amounts that he over-optimistically predicted.

PLANET EARTH CALLING ...

Marconi moved onto Tesla's patch again when he claimed to have detected radio signals coming from outside the atmosphere. Tesla pooh-poohed this, claiming that Marconi was only picking up signals from other terrestrial wireless operators. This could not have been the case with his own signals from Mars in 1899, he claimed, as there were no radio transmitters with a range of more than a few miles at that time. Nevertheless Robert Johnson wrote to Tesla pointing out that, once again, when Marconi had repeated one of his ideas, it was 'no longer laughed at'.

'Communication with intelligence on other stars? It may someday be possible,' said Marconi. Language may be a problem, but Marconi said:

Well, it is an obstacle, but I don't think it is insurmountable. You see, one might get through some message such as 2 plus 2 equals 4, and go on repeating it until an answer came back signifying 'Yes', which would be one word. Mathematics must be the same throughout the physical universe. By sticking to mathematics over a number of years one might come to speech. It is certainly possible.[22]

Tesla said that he had little confidence in Marconi's idea of trying to communicate with aliens using mathematics. He thought it would be better to send pictures by wireless - the human face, for example. *The New York Times* was surely being sarcastic when it suggested that they follow up by sending movies. Perhaps it could even be a commercial enterprise. 'With this beginning whole feature films can be sent by radio across the solar system and released on Mars on the night that sees their premiere on Broadway,'[23] the paper said.

In 1919, rocket-pioneer Robert Goddard (1882 - 1945) published *A Method of Reaching Extreme Altitudes,* claiming that it would be possible to send things as far as the Moon. Tesla said that this scheme seemed far-fetched as the fuels then known did not have the necessary 'explosive power'. He also doubted that any rocket could 'operate at 459° [F] below zero - the temperature of interplanetary space'.[24]

Fig.2.

PART FOUR

DESCENT AND RE-ASSESSMENT

CHAPTER 13

TALKING TO PIGEONS

I have been feeding pigeons, thousands of them, for years; thousands of them. But there was one pigeon, a beautiful bird, pure white with light grey tips on its wings; that one was different. It was a female. I would know that pigeon anywhere. No matter where I was, that pigeon would find me; when I wanted her I had only to wish and call her and she would come flying to me. She understood me and I understood her. I loved that pigeon. I loved her as a man loves a woman, and she loved me … As long as I had her, there was a purpose in my life.

Nikola Tesla[1]

ack in New York, Tesla moved into the Hotel St Regis. Robert Johnson had been appointed ambassador to Italy, so his friends left for Rome. On his own, Tesla became more eccentric. He would circle the block three times before entering the hotel, avoiding stepping on the cracks in the sidewalk. And he was fanatical about cleanliness, except when it came to pigeons which he still fed outside the New York Public Library.

Having given up the appetites of his youth, he now practised what he called 'gastronomical frugality' to which he owed his perpetual youth. According to Hugo Gernsback, Tesla believed that most people not only eat all of their bodily ills, but actually ate themselves to death by either consuming too much, or else by eating food that does not agree with them.[2] His daily menu consisted of:

Breakfast: One to two pints of warm milk and a few eggs, prepared by himself.

Lunch: None whatsoever, as a rule.

Dinner: Celery soup or similar, a single piece of meat or fowl, potatoes and one other vegetable; a glass of light wine. For dessert, perhaps a slice of cheese, and invariably a big raw apple.

And that's all.[3]

While he ate very little, Tesla insisted that what he did eat must be of the highest quality. He was also an accomplished cook who invented a number of appetizing dishes.

'His only vice is his generosity,' said Gernsback. 'The man who, by the ignorant onlooker has often been called an idle dreamer, has made over a million dollars out of his inventions - and spent them as quickly on new ones. But Tesla is an idealist of the highest order and to such men, money itself means but little.'[4]

FILET MIGNON AND ROAST DUCK

Tesla's frugal diet in his later years is markedly different from his consumption in his heyday. Then at dinner he would enjoy thick steaks, preferably filet mignon,[5] and often two or three of them, though he never put on weight. He remained 10 stone (142 pounds or 64 kg), from 1888 to around 1926 except for a brief period of illness, when he lost 5 pounds (2 kg).

Later he turned to lamb, ordering a roast saddle large enough to serve several people. He would eat only the central portion of the tenderloin. Another favourite was a crown of baby lamb chops, or duck roasted under a layer of celery stalks - a dish of his own invention. He would often supervise its cooking in the kitchen and it would be the centre-piece when entertaining friends. But Tesla would only eat the meat either side of the breast bone.[6]

Gradually, he substituted boiled fish, then turned to a vegetarian diet. Throughout his life he drank milk and towards the end warm milk became the mainstay of his diet.[7] When he was young, he drank a lot of coffee. Although he decided it was bad for him, he had difficulty giving it up. So with each meal, he would order a pot of coffee so that he could smell the aroma. Eventually he went off the smell and gave up this practice. He had always avoided tea and cocoa, but, along with wine, he drank whisky. This, he believed, was responsible for the longevity of his ancestors and prolonged his own life. When Prohibition was introduced in 1919, Tesla denounced it, saying it was an intolerable infringement of the rights of an individual.[8] 'It imposes restrictions on the most needed and harmless of stimulants,' he said, 'while permitting unlimited consumption of poisons by all classes, from childhood to old age.'[9]

He admitted that he had consumed enough alcoholic beverages to 'form a lake

Nikola Tesla in 1920, aged 64.

of no mean dimensions'.[10] But, being a law-abiding citizen, he gave it up, declaring that it was 6 months before he could digest a meal and that abstinence would reduce his life-expectation to 130 years.[11]

'I feel sure that if everyone had done the same,' he said, 'millions of Americans would have shortened their life-span and thousands would have died in the first 2 years. A sudden change of diet or the omission of one of its important elements, especially in advanced years, is extremely dangerous.'[12]

In later years, after the repeal of Prohibition, he would have a bottle of wine brought in an ice bucket, but not have it opened. It remained there purely to show that he could restrain himself from drinking.

He had also been a heavy smoker in his youth, particularly enjoying cigars. However, when he was in his early twenties, one of his sisters fell ill. She said she could try to get better if he gave up smoking. She recovered and he never smoked again.[13]

EXTREME GERM PHOBIA

After studying microscopic organisms before he left Europe, Tesla developed a phobia about germs.[14] The washroom in his laboratory was private. No one else was allowed to use it. He would be impelled to wash his hands on the slightest pretext, insisting that his secretary provide a freshly laundered towel each time to dry them. He would also avoid shaking hands, usually keeping his hands clasped behind him in social situations.

This led to embarrassment when visitors advanced proffering their hand. If Tesla was caught unawares and his hand was shaken, he would rush to the washroom at the first possible opportunity to scrub it, ignoring any business the visitor was there to conduct. And he found it particularly nauseating when workmen ate their lunch with dirty hands.[15] Hotel staff were kept at a distance of at least 3 feet.

Head waiters grew used to his demand to be seated at a table that was not to be used by other customers. He needed a fresh table cloth with every meal and two dozen serviettes. The silverware had to be sterilized before it left the kitchen. Tesla would then pick each item up with a serviette, and polish it with another. Then he would drop both serviettes on the floor before attending to the next item of cutlery. And if a fly alighted on the table, he would insist that everything was removed from the table and the meal would start over.[16]

He lived in hotels that could meet his meticulous standards and only employed assistants that were scrupulously clean. When he visited the barbers to have his half-hour scalp massage three times a week, he insisted on fresh towels on his chair, but strangely he did not object to being shaved using the same brush and shaving mug as the other customers.[17]

STEPPING OUT IN STYLE

To the end of his life, Tesla was a fastidious dresser. Well-cut clothes suited his tall, slim figure. 'In the matter of clothes', he observed, 'the world takes a man at his own valuation.' He wore white monogrammed silk shirts that had to withstand constant laundering. Collars and cuffs were discarded after a single use, as were handkerchiefs. Ties were replaced every week. Costing a dollar each, the only colours he would consider were red and black. They were tied in the old-fashioned, four-in-hand style. His pyjamas also had his initials embroidered on the left breast and his linen arrived freshly packaged.

Except on formal occasions, Tesla wore high-laced shoes, possibly to give his ankles extra support because of his height. They extended halfway up his calves. He insisted on a long narrow shoe with a tapered, square toe which had to be handmade.

His suits had waisted coats and he usually wore a black bowler or derby, which gave him an air of quiet elegance. He carried a cane and wore grey suede gloves. At $2.50 a pair, they were also replaced weekly, even if they were as clean as they had been when they came from the makers.[18]

RESTLESS SLEEPING HABITS

Tesla claimed that he usually slept for just 2 hours a night, 3 hours being too much. But he would go to bed at 5 am and get up at 10, the extra three hours rest, he maintained was for quiet contemplation. Once a year, he would sleep for the full 5 hours which, he said, gave him a tremendous reserve of energy.

In this, he competed with Edison who claimed only to sleep 4 hours a night, though when he sat in his lab he would take two 3-hour naps a day. Tesla probably did the same. Hotel staff said that they often found him sitting transfixed and they found they could work around him in his home without disturbing him.[19]

He took brisk walks to aid his concentration, but even then he was often in a dream. People who he knew quite well could walk past him, even though he appeared to be looking straight at them. In 1935, he said he was lucky not to have been killed while jaywalking in such a state.[20] Two years later, he was hit by a taxi and badly injured though, refusing to see a doctor he limped home. He had three cracked ribs and was confined to bed for 6 months.

WARDENCLYFFE REVISITED

Tesla still had not given up on Wardenclyffe and sued the Waldorf-Astoria over its destruction. He maintained that he had

The demolition of Wardenclyffe, 1917.

put up Wardenclyffe as collateral against the $20,000 he owed. They assumed that it was theirs. They had torn it down to sell for scrap and resell the land. Tesla, of course, was expecting to make $30,000 a day from Wardenclyffe if it was ever finished. If he then paid the $20,000, the experimental station would then be his again. In court he insisted that the Waldorf-Astoria were supposed to take care of it but, even before it had been torn down, they had allowed vandals to break in and steal expensive equipment.

During the trial Tesla was to give a loving description of his lost palace. The attorney for the Waldorf-Astoria tried to block this testimony, but the judge let him go ahead.

The building formed a square about 100 ft by 100 ft [30 by 30 m]. It was divided into four compartments, with an office and a machine shop and two very large areas ... The engines were located on one side and the boilers on the other side, and in the centre, the chimney

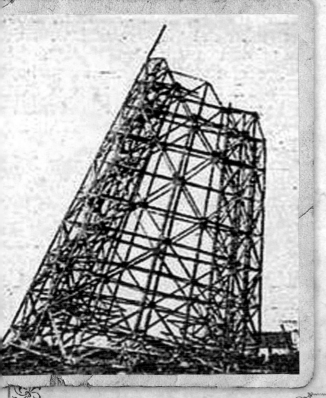

rose. There were two 300-horsepower boilers surrounded by two 16,000-gallon water tanks. To the right of the boiler plant was one 400-horsepower Westinghouse engine and a smaller 35-kilowatt engine to drive the dynamo for the lighting. Along with them was the main switchboard that controlled the pumps and various compressors.

Towards the road, on the railroad side, was the machine shop. That compartment was 100 ft by 35 ft [30 m by 10 m] with a door in the middle and it contained, I think eight lathes. Then there was the milling machine, a planer and shaper, a spliner, three drills, four motors, a grinder and a blacksmith's forge.

Now, in the compartment opposite, there was contained the real expensive apparatus. There were two special glass cases where I kept historical apparatus which was exhibited and described in my lectures and scientific articles. There were at least a thousand bulbs and tubes each of which represented a certain phase of scientific development.

Then there was also five large tanks, four of which contained special transformers created so as to transform the energy for the plant. They were about, I should say, 7 ft [2 m] high and about 5 ft by 5 ft [1.5 by 1.5 m] each, and were filled with special oil which we call transformer oil, to stand an electric tension of 60,000 volts. Then there was a fifth similar tank for special purposes. And there were my electric generating apparatus. That apparatus was precious because it could flash a message across the Atlantic, and yet it was built in 1894 or 1895.[21]

The Waldorf-Astoria's attorney objected again, but the judge allowed Tesla to continue.

Beyond the door of this compartment, there were to be condensers, what we call electric condensers, which would store the energy and then discharge and make it go around the world.

Some of these condensers were in an advanced state of construction, and others were not. Then there was a very expensive piece of apparatus that the Westinghouse Company furnished me, only two of its kind have ever been constructed. It was developed by myself with their engineers. That was a steel tank which contained a very elaborate assemblage of coils, and elaborate regulating apparatus, and it was intended to give every imaginable regulation that I wanted in my measurements and control of energy.[22]

He went on to describe a special 100-horsepower motor equipped with elaborate devices for rectifying the alternating currents and sending them back to the condensers. On this apparatus alone, Tesla said he had spent thousands of dollars. 'Then along the centre of the room I had a very precious piece of apparatus,' he said. It was his remote-controlled boat.

A BOLT FROM THE BLUE

Asked whether that was all, Tesla replied: 'Oh, no, nowhere near.' And he went on to describe a series of cupboards that contained numerous devices that each represented a different phase in the development of his work. In the testing room there were other instruments, some of which had been given to him by Lord Kelvin. Tesla then described the tower with its expensive underground workings containing special apparatus for 'gripping the earth'.[23]

The shaft, your honour, was first covered with timber and the inside with steel. In the centre of this there was a winding stairs going down and in the centre of the stairs was a big shaft through which the current was to pass, and this shaft was so figured in order to tell exactly where the nodal point is, so I could calculate exactly the size of the Earth or the diameter of the Earth and measure it exactly within 4 ft [1.2 m] with that machine.

And then the real expensive work was to connect that central part with the earth, and there I had special machines rigged up which would push the iron pipes, one length after another, and I pushed, I think 16 of them, 300 ft [90 m]. The current through these pipes was to take hold of the earth. Now that was a very expensive part of the work, but it does not show on the tower, but it belongs to the tower.

The primary purpose of the tower, your honour, was to telephone, to send the human voice and likeness around the globe. That was my discovery, that I announced in 1893, and now all the wireless plants are doing that. There is not another system being used. Then the idea was to reproduce this apparatus and connect it just with a central station and telephone office, so that you may pick up your telephone and if you wanted to talk to a subscriber in Australia you would simply call up that plant and that plant would connect you immediately. And I had contemplated to have press messages, stock quotations, pictures for the press and reproductions of signatures, cheques and everything transmitted from there.[24]

Tesla was asked whether he had any warning that the tower was going to be demolished. He replied, 'No, sir. It came like a bolt from a blue sky...'[25]

Although the Waldorf-Astoria had not accounted for the machinery, they had sold off and destroyed property worth $350,000 to recoup $20,000, the judge found in favour of the hotel, who also had the last word:

As a solace to the wild hopes of this dreamy inventor, if prior to that time he should grasp in his fingers any one of the castles in Spain which always were floating about in his dreams, and had he paid the board bills which he owed, this wild scrubby woodland, including the Tower of Babel thereon, would cheerfully have been reconveyed to him.[26]

LISTENING TO COMMUNIST OVERTURES

The 1920 US Presidential Election was the first to be broadcast to a national audience - though Lee De Forest had announced the wrong winner to a small audience four years before. In Italy, Fascist leader Benito Mussolini (1883 - 1945) saluted Marconi. His national broadcasting system allowed the dictator to reach the entire Italian nation after he came to power in 1922. Meanwhile, Vladimir Ilich Lenin (1870 - 1924), who had led the Bolshevik Revolution in Russia in 1917, made overtures to Tesla, asking him to come to the Soviet Union to build power stations and an AC distribution system.

Tesla was also invited to speak at a meeting of the Friends of Soviet Russia in Springfield, Massachusetts, travelling there with Ivan Mashevkief from the Russian Workers Club of Manhattan. At the meeting, the organizers, a group of Italian radicals, addressed Tesla mysteriously as 'Bettini'. According to an FBI agent at the event, Tesla 'prophesied that Italy would soon adopt a communist form of government'.[27] However, there is no evidence that he knew what he was getting himself into. Tesla took little interest in politics and he was, at best, naïve.

WAITING FOR THE MIDNIGHT HOUR

While commuting to Milwaukee, Tesla was paying $15 a day for room 1607 in the Hotel St Regis on Fifth Avenue. However, he neglected to pay for 7 months and was forced to move to the Hotel Marguery on Park Avenue and 48th Street. Fortunately this was still close to his night-time haunts, Grand Central Station and Bryant Park behind New York Public Library. He was spotted there one night, sporting his derby, cane and white gloves by the *New York World*:

Midnight is the hour he chooses for his visits ... Tall, well-dressed, of dignified bearing, he whistles several times, a signal for the pigeons on the ledges of the building to flutter down about his feet. With a generous hand, the man scatters peanuts on the lawn from a bag. A proud man, yet humble in his charities – Nikola Tesla.[28]

HOW THIS MAN WORKED ...

A young science journalist named Kenneth Swezey, once praised by Albert Einstein for his explanation of the Archimedes principle, sought out Tesla. While Swezey was only 19 and Tesla in his late sixties, they became firm friends for the rest of his life. According to Swezey, Tesla sometimes greeted him at the door stark naked, but insisted that Tesla was 'absolutely celibate'.

He confirmed that Tesla slept less than 2 hours a night, though he would occasionally doze to recharge his batteries. He would walk 8 to 10 miles a day and relax in a bathtub, though gone was the electric shower that bombarded him with cleansing particles. He would also clench and unclench his toes a hundred times each night to stimulate the brain cells.[29] Swezey recalled:

And how this man worked! I will tell you about a little episode ... I was sleeping in my room like one dead. It was three after midnight. Suddenly the telephone ring awakened me. Through my sleep I heard his voice, 'Swezey, how are you, what are you doing?' This was one of many conversations in which I did not succeed in participating. He spoke animatedly, with pauses [as he had worked] out a problem, comparing one theory to another, commenting: and when he felt he arrived at the solution, he suddenly closed the telephone.[30]

In 1926, Tesla moved to the Hotel Pennsylvania. There the 'tall, thin, ascetic man' was interviewed by *Colliers* magazine and he gave another of his predictions:

'This struggle of the human female toward sex equality will end in a new social order, with the human female as superior.'[31] Tesla forwarded his prediction to J.P. Morgan's daughter Anne, who he had remained in touch with and was now an advocate for the women's movement.

INTO THE REALMS OF THE FUTURE

It was around that time, Tesla retired. Miss Dorothy F. Skerritt, Tesla's secretary until he closed his office at 70, described him at that age:

As one approached Mr Tesla you beheld a tall, gaunt man. He appeared to be an almost divine being. When about 70, he stood erect, his extremely thin body immaculately and simply attired in clothing of a subdued colouring. Neither scarf pin nor ring adorned him. His bushy black hair was parted in the middle and brushed back briskly from his high broad forehead, deeply lined by his close concentration on scientific problems that stimulated and fascinated him. From under protruding eyebrows his deep-set, steel grey, soft, yet piercing eyes, seemed to read your innermost thoughts. As he waxed enthusiastic about fields to conquer and achievements to attain his face glowed with almost ethereal radiance, and his listeners were transported from the commonplaces of today to imaginative realms of the future. His genial smile and nobility of bearing always denoted the gentlemanly characteristics that were so ingrained in his soul.[32]

TOO GREAT A SACRIFICE

In an interview with the *New York World* in 1926, he said: 'Sometimes I feel that by not marrying I made too great a sacrifice to my work, so I have decided to lavish all the affection of a man no longer young on the feathery tribe. I am satisfied if anything I do will live for posterity. But to care for those homeless, hungry or sick birds is the delights of my life. It is my only means of playing.'[33]

One particular pigeon was special to him. When he found it, it had a broken leg and wing. 'Using my mechanical knowledge, I invented a device by which I supported its body in comfort in order to let the bones heal.'[34] He kept the bird in his hotel suite and figured that it cost him more than $2,000 to heal her. It took a year-and-a-half before she was well again and, Tesla said, she was 'one of the finest and prettiest birds I have ever seen'.[35]

MY LIFE'S WORK WAS FINISHED ...

His love of that one pigeon and her death affected him profoundly. He told John J. O'Neill:

When she was ill I knew, and understood; she came to my room and I stayed beside her for days. I nursed her back to health. That pigeon was the joy of my life. If she needed me, nothing else mattered. As long as I had her, there was a purpose in my life.

Then one night as I was lying in my bed in the dark, solving problems, as usual, she flew in through the open window and stood on my desk. I knew she wanted me; she wanted to tell me something important so I got up and went to her.

As I looked at her I knew she wanted to tell me – she was dying. And then, as I got her message, there came a light from her eyes – powerful beams of light ... it was a real light, a powerful, dazzling, blinding light, a light more intense than I had ever produced by the most powerful lamps in my laboratory.

When that pigeon died, something went out of my life. Up to that time I knew with a certainty that I would complete my work, no matter how ambitious my programme, but when that something went out of my life I knew my life's work was finished.[36]

Nikola Tesla in 1933, aged 77.

OCT 13

CHAPTER 14

A NEW SOURCE OF ENERGY

I'm happy to hear that you are celebrating your 75th birthday, and that, as a successful pioneer in the field of high-frequency currents, you have been able to witness the wonderful development of this field of technology. I congratulate you on the magnificent success of your life's work.

Albert Einstein[1]

While working on his petrol-powered turbine in Philadelphia in 1924 - 25, Tesla met John B. Flowers, an inspector of an aircraft factory. With development of his bladeless turbine reaching a dead end, Tesla returned to the idea of powering planes and cars remotely from large central power stations like the one he had tried to build at Wardenclyffe. In Tesla's hotel suite, Flowers helped draft a proposal to test and implement Tesla's Wireless Power System to present it to J. H. Dillinger, head of the Radio Laboratory at the Bureau of Standards in Washington, DC.

Flowers told Dillinger that Tesla's system would power a plane at any point around the world and that Tesla had already developed an oscillator to provide the power which he was willing to give to the American government if they agreed to build the plant. A meeting in Washington was arranged and Dillinger sent the proposal to physicist Harvey L. Curtis (1875 - 1956).[2]

The 10-page document outlined a plan to use standing waves to operate cars and planes. To demonstrate this, a sketch showed a balloon, standing in for the Earth, and a mechanical oscillator standing in for the electrical device:

... a mechanical oscillator arm was fastened to the tied opening of a rubber balloon 20 inches [50 cm] in diameter. The oscillator arm was operated with an electrical motor at 1,750 rpm by means of an eccentric on the motor shaft. The balloon hung free in the air. The rubber surface of the balloon represented the earth's conducting surface and the air inside its insulating interior. The waves were propagated in the rubber surface at the rate of 51 ft per second [15.5 m per second], the frequency of transmission was 29 cycles per second and the wavelength was 21 inches [53 cm].

The mechanical oscillator was used in place of Tesla's electrical oscillator as it presents an almost perfect analogy. Standing or stationary waves of the rubber surface replace the electromagnetic waves of Tesla's system. By the test of this analogue, the operation of Tesla's system can be forecast. When the oscillator arm was set in motion by operating the motor, there were three standing waves having six loops on the Earth's surface – all having the same amplitude of vibration!

When the finger was pushed against one or more loops, all the loops were reduced in amplitude in the same proportion showing the ability to obtain all the power out at one or more points! The waves extended completely around the world and returned to the sending station.[3]

Curtis rejected the proposal as, with Tesla's standing waves, there would be a concentration of energy at the nodes. But, as Curtis pointed out:

The system proposed by Mr Flowers does not have this feature. He proposes to collect energy at any point ... some means would have to be devised for concentrating this energy and making it available. No such method has been proposed, and I do not think of any that would be feasible ... I do not know of any wireless apparatus of sufficient magnitude to warrant the expectation that power can be economically transmitted by radio methods.[4]

FASTER THAN THE SPEED OF LIGHT

Tesla denied that the electricity would be available only at the nodal points, advising that, in a hydraulic system, the pressure of the fluid is the same throughout. There was energy available at an electrical outlet even when nothing is plugged in. He explained that the oscillations would spread from his magnifying transmitter theoretically at an

ALBERT EINSTEIN (1879 – 1955)

Born in Germany, Einstein completed his education in Switzerland and took a job in the patent office in Bern. It was there he realized that, if the speed of light was an absolute, Isaac Newton's laws of motion must be rewritten to accommodate Maxwell's equations. The result was his theory of special relativity, which he published in 1905. He then realized that this did not deal with acceleration or gravity. He combined those to make his theory of general relativity in 1915. This was confirmed by astronomical observation 4 years later and he was awarded the Nobel Prize for physics in 1921.

In 1933, he emigrated to the United States, settling in Princeton. Einstein's equation $E=mc^2$ predicts the possibility of making an atomic bomb. Fearing that Nazi Germany may well have been on their way to doing so, Einstein was persuaded to sign a letter to President Franklin D. Roosevelt warning him of the possibility. He spent the rest of his life trying to develop a unified field theory that would combine all the forces of motion into one theoretical framework.

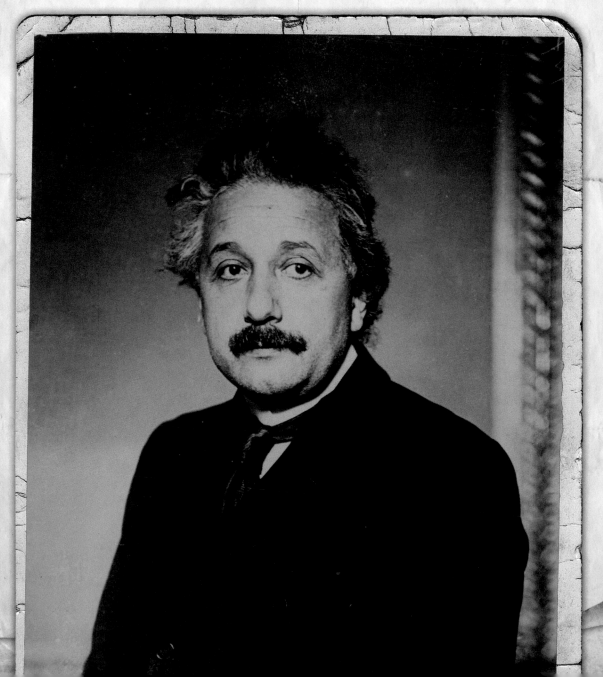

infinite speed, then slow down - at first very quickly, then at slower rate. After around 6,000 miles (9,500 km) they would travel at the speed of light.

> *From there on it increases in speed, slowly at first, and then more rapidly, reaching the antipode with approximately an infinite velocity. The laws of motion can be expressed by stating that the waves on the terrestrial surface sweep in equal intervals of time over equal areas, but it must be understood that the current penetrates deep into the Earth and the effects produced on the receivers are the same as if the whole flow was confined to the Earth's axis joining the transmitter with the antipode. The mean surface speed is thus about 471,200 km per second [292,790 miles a second] – 57 per cent greater than that of the so-called Hertz waves.[5]*

There was a problem with this. James Clerk Maxwell's equations predicted the speed of light at 186,000 miles a second (300,000 km a second). At first, it was assumed that this speed was relative to the background ether that electromagnetic radiation propagated through. But the Michelson-Morley experiment showed there was no such thing as ether. Einstein realized that this meant the speed of light was an absolute - and there was no such thing as a speed faster than light.

Tesla railed against Einstein and relativity. He would not accept the concept of curved space, as predicted by Einstein's theory of general relativity, either. But Einstein, with his Nobel Prize, was the new star in the scientific firmament. Tesla, Edison, who died in 1931, Bell, who died in 1922, and the Wright brothers, who died in 1912 and 1948, were old hat.

BUSINESS AS USUAL

Despite the rejection of their plans in Washington, Flowers and Tesla went to Detroit to try and sell his 'flying automobile' to General Motors. Tesla also tried to sell his speedometer to Ford, but its high cost made it better suited to luxury cars. In Detroit, Tesla met his nephew Nicholas Trbojevich and there was an incident that became part of family lore. The two were going for a late snack in an expensive hotel. The head waiter suggested that they wait five minutes. Then the $5 cover charge would be lifted. This was in the middle of the Great Depression when $5 would feed a family for a week. But Tesla was not prepared to wait. When Trbojevich questioned his uncle over the matter of the cover charge, Tesla said: 'I'll never die rich unless the money comes in the door faster than I can shovel it out of the window.'[6]

Tesla held talks with US Steel concerning installing his bladeless turbines on the exhaust from the blast furnaces, generating huge amounts of electricity. But, apparently, a test did not go ahead. Then in Buffalo, Tesla conducted some top-secret experiments. It was said that the petrol engine of a Pierce-Arrow sedan was replaced by an AC induction motor. A 'power receiver' using 12 vacuum tubes was set in the dashboard connected to a 6 ft (2 m) antenna.[7] There is other speculation that it was powered by a steam or petrol-driven turbine, but no physical evidence of either design has been found.

While experimenters were using Tesla Coils to try and split the atom, Tesla himself was making more outlandish predictions, saying that all the machinery on Earth could be powered by cosmic rays. Unlimited quantities of power could be transmitted through wires or wirelessly from a central station to anywhere on the globe,

Left: Albert Einstein in 1921, the year he won the Nobel Prize for physics.

Tesla examining electrical apparatus in his laboratory.

eliminating the need for coal, oil, gas or any other terrestrial energy source. Already the central source of energy on Earth was the Sun, he said, but the new source of power would not be turned off at night.[8]

HERALDING A NEW INDUSTRIAL REVOLUTION

With just five days to go to his 75th birthday, Tesla said that he would soon announce 'by far the most important discovery' of his long career. 'It will throw light on many puzzling phenomena of the cosmos,' he said, 'and may prove also of great industrial value, particularly in creating a new and virtually unlimited market for steel.'[9]

He said that he had been wonderfully fortunate in coming up with new ideas that he was sure would be remembered by posterity. He was confident that his rotating magnetic field, induction motor and wireless system would live on long after he was gone, but he still considered his latest discoveries the most important. They would mark a new departure in science, be of great practical values and inaugurate a new industrial revolution, he said.

He had already succeeded in proving his theories by experimentation and, if the calculations based on them turned out to be true, the world would have a new source of energy in practically unlimited amounts, available at any point on the globe. But, again, he was tantalizingly vague when it came to the details:

I can only say at this time that it will come from an entirely new and unsuspected source, and will be for all practical purposes constant, day and night, and at all times of the year. The apparatus for capturing the energy and transforming it will partake both mechanical and electrical features, and will be of ideal simplicity. At first the cost may be found

too high, but this obstacle will be overcome. Moreover, the installment will be, so to speak, indestructible, and will continue to function for any length of time without additional expenditures.[10]

The press had heard such promises from Tesla before and wanted to know when he was going to make an official announcement of his new discoveries. But the great man was unwilling to be pinned down. These ideas had not come to him overnight, but as the result of intense study and experimentation for nearly 36 years. He said he was anxious to give the facts to the world as soon as possible, but wished to present them in a finished form. That may take a few months, or a few years, he said.

All the energy that the Earth receives from all the suns and stars of the universe is only about one-quarter of one per cent of that which it receives directly from the Sun. Therefore, it would be incomparably more rational to harness the heat and light rays of the Sun than attempt to capture the insignificant energy of this radiation ... We can do it now, and we are doing it to a certain extent. But the tremendous handicap is found in the periodic character of this kind of energy supply. Many attempts have been made in this direction, but invariably it was found that the power was too expensive.[11]

DISMISSING ATOMIC ENERGY

Having rejected Einstein's theory of relativity, Tesla also dismissed the idea of atomic energy. 'The idea of atomic energy is illusionary,' he said, 'but it has taken so powerful a hold on the minds that, although I have preached against it for 25 years, there still are some who believe it to be realizable.'[12]

He claimed to have disintegrated atoms in his experiments with the high-potential vacuum tube he developed in 1896, which he considered one of his best inventions. He operated at a range of potentials from 4 million to 18 millions volts. More recently, he said, he had designed an apparatus that would work at 50 million volts, which should produce results of great scientific importance. 'But as to atomic energy, my experimental observations have shown that the process of disintegration is not accompanied by a liberation of such energy as might be expected from present theories,'[13] he said.

COSMIC RAYS AND BEYOND

Tesla claims to have discovered cosmic rays while investigating X-rays and radioactivity in Colorado Springs in 1899, but his findings were in disagreement with theories advanced more recently:

I have satisfied myself that the rays are not generated by the formation of new matter in space, a process which would be like water running up hill. Nor do they come to any appreciable amount from the stars. According to my investigations the Sun emits a radiation of such a penetrative power that it is virtually impossible to absorb it in lead or other substances. It has, furthermore, other extraordinary properties in regard to which I shall express myself at some future date. This ray, which I call the primary solar ray, gives rise to a secondary radiation by impact against the cosmic dust scattered through space. It is the secondary radiation which now is commonly called the cosmic rays, and comes, of course, equally from all directions in space.[14]

He also dismissed the idea that radioactivity resulted from activity within radioactive substances. It was caused by rays emitted from the Sun. If radium could be screened effectively from this ray, it would cease to be radioactive, he said. He had also been designing rocket-ships that he said could

attain speeds of nearly a mile a second – 3,600 miles an hour (5,793 km per hour) – through the rarefied medium above the stratosphere. Again, he hoped his rocket-ships would bring world peace:

I anticipate that such machines will be of tremendous importance in international conflicts in the future. I foresee that in times not too distant wars between various countries will be carried on without a single combatant passing the border. At this very time it is possible to construct such infernal machines which will

Nikola Tesla appeared on the cover of *Time* magazine, 20 July 1931.

carry any desired quantity of poisoned gases and explosives, launch them against a target thousands of miles away and destroy a whole city. If wars are not done away with, we are bound to come eventually to this kind of warfare, because it is the most economical means of inflicting injury and striking terror in the hearts of enemies that ever has been imagined. Densely populated countries, like England and Japan, will be at a great disadvantage as compared with those embracing vast territories, such as the United States and Russia.[15]

Although some of Tesla's ideas in later life can be dismissed as the ravings of a mad scientist, sometimes he shows remarkable prescience.

SENDING SIGNALS TO THE STARS

When *Time* magazine put the ageing and eccentric inventor on their cover of the 20 July 1931 issue to celebrate his 75th birthday, Tesla did not disappoint. He told them of his new invention, the Tesla-scope, that he could use to signal to the stars, saying:

I think that nothing can be more important than interplanetary communication. It will certainly come some day, and the certitude that there are other human beings in the universe, working, suffering, struggling like ourselves, will produce a magic effect on mankind, and will form the foundation of a universal brotherhood that will last as long as humanity itself.[16]

Asked when this would happen, he said: 'I have been leading a secluded life, one of continuous, concentrated thought and deep meditation. Naturally enough I have accumulated a great number of ideas. The question is whether my physical powers will be adequate to working them out and giving them to the world.'

He also claimed that with 15 million volts - 'the highest ever used' - he split atoms

over and over again, but no energy was released. The *Time* article carried several other amusing stories about Tesla. It said that he left the 'swank' Hotel St Regis after the maids complained that he kept four pet pigeons in his roll-top desk and that, while walking down an icy Fifth Avenue, 'he slipped, threw himself into a flying somersault, landed on his feet, unperturbed kept on walking'. *Time* reported more eccentricities:

At the Hotel Governor Clinton where he now lives, if someone rings him up on the telephone or knocks at his door and he does not want to answer, he locks himself in the bathroom, turns the water loudly on. He is very sensitive to sensory stimuli. When he gets excited, blinding lights flash through his mind. He retreats to bed. A lifelong bachelor, habitually he goes to bed at 5.30 am, rises at 10.30 am. But he does not sleep the whole period. Proudly, yet almost plaintively, he explains: 'I roll around and work on my problems.'[17]

THE PLAUDITS OF PEERS

Birthday accolades flooded in. Over a hundred letters of congratulation were received from other scientists or inventors including Sir Oliver Lodge, Lee De Forest and Albert Einstein. Notably absent were birthday greetings from Marconi and Pupin. And none could match the tribute bestowed on Edison when he died 3 months later and the lights of New York were dimmed in reverence.

Naturally, Tesla was full of new predictions. 'I feel that we are nearing a period when the human mind will perform greater wonders than ever before,' he said. 'This is due to the continuous refinement of means and methods of observation and ever-increasing delicacy of our perception.'[18]

We were about to conquer nature, contact

beings on other planets and transmit huge amounts of power vast distances. A reporter from *The New York Times* again asked when he was going to make his discovery public. 'There was a trace of regret in his voice as he answered,' said the paper, 'and the look of a man who has work enough for centuries and only a few years to do it in.'[19] Tesla then quoted Goethe. 'He had not read Goethe for 40 years, he said, and he quoted it from memory.'[20]

He was also at odds with the new ideas of quantum mechanics. 'There is no chance in nature,' he said, 'although the modern theory of indeterminacy attempts to show scientifically that events are governed by chance. I positively deny that. The causes and effects, however complex, are intimately linked, and the result of all inferences must be inevitably fixed as by a mathematical formula.'[21]

WAVES IN TIME AND SPACE

He re-asserted that human beings were automatons completely under the control of external forces and he denied the existence of individuality, saying:

It took me not less than 20 years to develop a faculty to trace every thought or act of mine to an external influence. We are just waves in time and space, changing continuously, and the illusion of individuality is produced through the concatenation of the rapidly succeeding phases of existence. What we define as likeness is merely the result of the symmetrical arrangement of molecules which compose our body.[22]

He also denied the existence of the soul or spirit, saying they were merely expressions of the functions of the body. 'These functions cease with death and so do soul and spirit,' he said. 'What humanity needs is ideals. Idealism is the force that will free us from material fetters.'[23]

PHOTOGRAPHING THOUGHT

At 77, Tesla told a journalist from the *Kansas City Journal-Post* that he expected soon to be able to photograph thoughts, explaining:

In 1893, while engaged in certain investigations, I became convinced that a definite image formed in thought must, by reflex action, produce a corresponding image on the retina, which might possibly be read by suitable apparatus. This brought me to my system of television, which I announced at that time. My idea was to employ an artificial retina receiving the image of the object seen, an 'optic nerve' and another such retina at the place of reproduction. These two retinas were to be constructed after the fashion of a checkerboard with many separate little sections, and the so-called optic nerve was nothing more than a part of the earth.

An invention of mine enables me to transmit simultaneously, and without any interference whatsoever, hundreds of thousands of distinct impulses through the ground just as though I had so many separate wires. I did not contemplate using any moving part — a scanning apparatus or a cathodic ray, which is a sort of moving device, the use of which I suggested in one of my lectures.

Now if it be true that a thought reflects an image on the retina, it is a mere question of illuminating the same property and taking photographs, and then using the ordinary methods which are available to project the image on a screen. If this can be done successfully, then the objects imagined by a person would be clearly reflected on the screen as they are formed, and in this way every thought of the individual could be read. Our minds would then, indeed, be like open books.[24]

As always, he claimed to have discovered a new source of power. He was not ready to go into details. He had to check his findings before they could be formally announced.

But he had been working the underlying principles for many years. From the practical point of view, his generator would require a huge initial investment, but once a machine was installed it would work indefinitely and the cost of operation would be next to nothing.

But this time he gave more details. The design was relatively simple - 'just a big mass of steel, copper and aluminium, comprising a stationary and rotating part, peculiarly assembled'.[25] The electricity would then be distributed long distances by his AC system which, he said, already distributed 30 million horsepower of waterpower, and there were projects then going on all over the world which would double that amount.

'Unfortunately, there is not enough water power to satisfy the present needs,' he said, 'and everywhere inventors and engineers are endeavouring to unlock some additional store of energy.'[26]

THE FORMULA FOR A LONG LIFE

In his eighth decade, Tesla still expected to live a long time and reflected on life and longevity.

Quite early in life I set about disciplining myself, planning out a programme of living for what I considered the most sane and worthwhile life. Since I love my work above all things, it is only natural that I should wish to continue it until I die. I want no vacation – no surcease from my labours. If people would select a life work compatible with their temperaments, the sum total of happiness would be immeasurably increased in the world.

Many are saddened and depressed by the brevity of life. 'What is the use of attempting to accomplish anything?' they say. 'Life is so short. We may never live to see the completion of the task.' Well, people could prolong their lives considerably if they would but make the effort. Human beings do so many things that pave the way to an early grave.

Above: Tesla's theoretical invention, the thought camera, projecting human thoughts onto a screen.

First of all, we eat too much, but this we have heard said often before. And we eat the wrong kinds of foods and drink the wrong kinds of liquids. Most of the harm is done by overeating and under-exercising, which bring about toxic conditions in the body and make it impossible to throw off the accumulated poisons.

My regime for the good life and my diet? Well, for one thing, I drink plenty of milk and water. Why overburden the bodies that serve us? I eat but two meals a day, and I avoid all acid-producing foods. Almost everyone eats too many peas and beans and other foods containing uric acid and other poisons. I partake liberally of fresh vegetables, fish and meat sparingly, and rarely. Fish is reputed as fine brain food, but has a very strong acid reaction, as it contains a great deal of phosphorus. Acidity is by far the worst enemy to fight off in old age.

Potatoes are splendid, and should be eaten at least once a day. They contain valuable mineral salts and are neutralizing. I believe in plenty of exercise. I walk 8 or 10 miles every day, and never take a cab or other conveyances when I have the time to use leg power. I also exercise in my bath daily, for I think that this is of great importance. I take a warm bath, followed by a prolonged cold shower.

Sleep? I scarcely ever sleep. I come of a long-lived family, but it is noted for its poor sleepers. I expect to match the records of my ancestors and live to be at least 100. My sleeplessness does not worry me. Sometimes I doze for an hour or so. Occasionally, however, once in a few months, I may sleep for 4 or 5 hours. Then I awaken virtually charged with energy, like a battery. Nothing can stop me after such a night. I feel great strength then. There is no doubt about it but that sleep is a restorer, a vitalizer, that it increases energy. But on the other hand, I do not think it is essential to one's well being, particularly if one is habitually a poor sleeper.

Today, at 77, as a result of a well-regulated life, sleeplessness notwithstanding, I have an excellent certificate of health. I never felt better in my life. I am energetic, strong, in full possession of all my mental facilities. In my prime I did not possess the energy I have today. And what is more, in solving my problems I use but a small part of the energy I possess, for I have learned how to conserve it. Because of my experience and knowledge gained through the years, my tasks are much lighter. Contrary to general belief, work comes easier for older people if they are in good health, because they have learned through years of practice how to arrive at a given place by the shortest path.[27]

DEVELOPING THE DEATH RAY

Tesla had inherited a deep hatred of war from his father. Throughout his life, he sought ways to end warfare. Short of that, he thought wars should be fought out between machines. His idea for a death ray began back in the 1890s when he produced a type of lamp which, with a beam of electrons, could vaporize zirconia or diamonds. And in 1915, he talked of beaming energy from Wardenclyffe, that would 'paralyze or kill'.

In World War I, British inventor Harry Grindell-Matthews claimed to have invented a 'diabolical ray' that could be used against zeppelins. In the early 1920s, both the British and the French governments showed an interest. In 1924, he went to New York where he met Hugo Gernsback and, probably, Tesla. However, Gernsback and Professor W. Severinghouse, a physicist from Columbia University, tried unsuccessfully to duplicate his findings. Not to be outdone, the Germans and the Soviets both claimed to have developed beams that could bring down planes.

But, Tesla was not convinced. In 1934, he said: 'It is impossible to develop such a ray. I worked on that idea for many years before my ignorance was dispelled and I became convinced that it could not be realized.'

He was working on something that he said was entirely different. 'This new beam of mine consists of minute bullets moving at a terrific speed, and any amount of power desired can be transmitted by them. The whole plant is just a gun, but one which is incomparably superior to the present.'[28]

The war clouds were gathering over Europe again and, on 11 July 1934, *The New York Times* carried the headline on its front page, reading: *Tesla, at 78, Bares New 'Death Beam'*. Tesla said his new invention 'will send concentrated beams of particles through the free air, of such tremendous energy that they will bring down a fleet of 10,000 enemy airplanes at a distance of 250 miles (400 km) from a defending nation's borders and will cause armies of millions to drop dead in their tracks'.[29]

The death beam would operate silently at distances as far as you could see with a telescope and limited only by the curvature of the Earth. It would be invisible and leave no marks beyond the evidence of its destruction. An army of a million men would be annihilated in a second and, even with the most powerful microscope, it would not be possible to see what had caused their deaths.

It would also be the perfect defence against bombing. 'The flying machine has completely demoralized the world,' he wrote, 'so much that in some cities, as London and Paris, people are in mortal fear of aerial bombing. The new means I have perfected afford absolute protection against this and other forms of attack.'[30]

Tesla said that his death beam would make war impossible by offering every country an 'invisible Chinese Wall, only a million times more impenetrable'.[31] It would make every nation impregnable to attack by aeroplanes or large invading armies.

While making every nation safe from invasion, Tesla said they could not be used as offensive weapons as the death beam 'could be generated only from large, stationary and immovable power plants, stationed in the manner of old-time forts at various strategic distances from each country's border ... they could not be moved for purposes of attack'.

However, he admitted that smaller generating plants could be mounted on battleships with enough power to destroy incoming aircraft - re-establishing the superiority of the battleship over the aeroplane again. Submarines could also become obsolete, he said, as methods of detecting them had been perfected to the point where there was no point in submerging. Once a submarine had been located, the death beam could be employed as it would work underwater, though not as well as in air.

Elsewhere he proclaimed that the battleship was doomed. 'What happened to the armoured knight will also happen to the armoured vessel,' he said. The money spent on battleships 'should be directed in channels that will improve the welfare of the country'.[32]

MANIFESTATIONS OF ENERGY

The production of the death beam involved four new inventions of Tesla's, though he would not provide details of these until they had been submitted to the proper scientific authorities. However, he said, the first invention was an apparatus for producing rays and 'other manifestations of energy in free air', eliminating the high vacuum necessary at present for the production of such rays and beams. The second was a new method for producing a very great electrical force. The third was a method for amplifying this and the fourth, he said, was 'a new method for producing a tremendous electrical repelling force'. Again Tesla was looking at a potential of 50 million volts which would catapult microscopic particles of matter towards the target.[33] He reckoned that it would cost no more than $2 million and take only three months to build.

'All my inventions are at the service of the United States government,'[34] he said.

Should the government take him up on his offer, he said he would go to work at once and keep on going until he collapsed. However, he added: 'I would have to insist on one condition - I would not suffer interference from any experts. They would have to trust me.'[35]

In the *New York Herald* journalist Joseph Alsop described the progress Tesla was making developing his death ray:

He illustrated the sort of thing that the particles will be by recalling an incident that occurred often enough when he was experimenting with a cathode tube. Then, sometimes, a particle larger than an electron, but still very tiny, would break off from the cathode, pass out of the tube and hit him. He said he could feel a sharp, stinging pain where it entered his body, and again at the place where it passed out. The particles in the beam of force, ammunition which the operators of the generating machine will have to supply, will travel far faster than such particles as broke off from the cathode, and they will travel in concentrations, he said.

As Dr Tesla explained it, the tremendous speed of the particles will give them their destruction-dealing qualities. All but the thickest armoured surfaces confronting them would be melted through in an instant by the heat generated in the concussion. Such beams or rays of particles now known to science are composed always of fragments of atoms, whereas, according to Dr Tesla, his would be of microscopic dust of a suitable sort.

The chief differentiation between his and the present rays would appear to be, however, that his are produced in free air instead of in a vacuum tube. The vacuum tube rays have been projected out into the air, but there they travel only a few inches, and they are capable only of causing burns or slight disintegration of objects which they strike.[36]

Tesla tried to get Jack Morgan to finance a prototype of his invention, but Morgan was unconvinced. He tried to deal directly with the British, to no avail. Frustrated, he sent an elaborate technical paper, including diagrams, to a number of nations including the US, Canada, Britain, France, the Soviet Union and Yugoslavia. Called 'The New Art of Projecting Concentrated Non-Dispersive Energy Through Natural Media', the paper provided the first technical description of a charged particle beam weapon. And it was not just fantasy. Tesla had solved one of the key problems of a death ray - how to operate a vacuum chamber with one end open to the atmosphere. He achieved this by directing a high-velocity stream of air at the tip of his gun to maintain 'dynamic seal'. This would be provided by a large Tesla turbine.[37]

Interest came from the Soviet Union and, in 1937, Tesla presented a plan to the Amtorg Trading Corporation, in New York City, which handled trade with the Soviet Union. Two years later, in 1939, part of the prototype was tested in the USSR and Tesla received a payment of $25,000. But by then, the Soviet Union had allied itself with Nazi Germany.[38]

While Tesla's death beam did not see the light of day in World War II, during the Cold War both the US and the Soviet Union worked on charged particle beams.

RESEMBLING DR FRANKENSTEIN

Tesla made a further move into science fiction when the 1931 horror classic *Frankenstein* used Tesla Coils to make lightning flashes. Much of the equipment used by Dr Frankenstein bears an uncanny resemblance to the apparatus Tesla invented for his experiments. Indeed, Tesla favoured the movie's producer Carl Laemmle as he fought patent battles with Edison when establishing Universal Pictures.[39]

In 1935, one of Tesla's electrical extravaganzas was filmed by a newsreel camera man and offered to Paramount, but they found the subject too technical. Nevertheless Hugo Gernsback and Frank Paul continued to use Tesla's ideas in their sci-fi comics.[40]

Meanwhile Tesla went about work on his death ray in a secret laboratory under the Fifty-ninth Street Bridge. One of his other inventions of the period was a wooden birdcage, complete with birdbath. Western Union boys were despatched with these to rescue injured pigeons from around New York Public Library, Bryant Park and St Patrick's Cathedral.[41]

TAINTED WITH ANTI-SEMITISM

Tesla also had ties with a Hungarian architect named Titus deBobula, possibly through the Puskás brothers. deBobula borrowed money from the inventor as early as 1900. In 1908, he married the niece of Pittsburgh steel magnate Charles Schwab (1862 - 1939). deBobula then designed and built Schwab's new mansion and borrowed money from him for a number of real estate ventures. He offered to find the backing to rebuild Wardenclyffe, but deBobula's ventures turned sour. He fell out with Schwab and became an anarchist. Back in Budapest, he joined a pro-Hitler group and

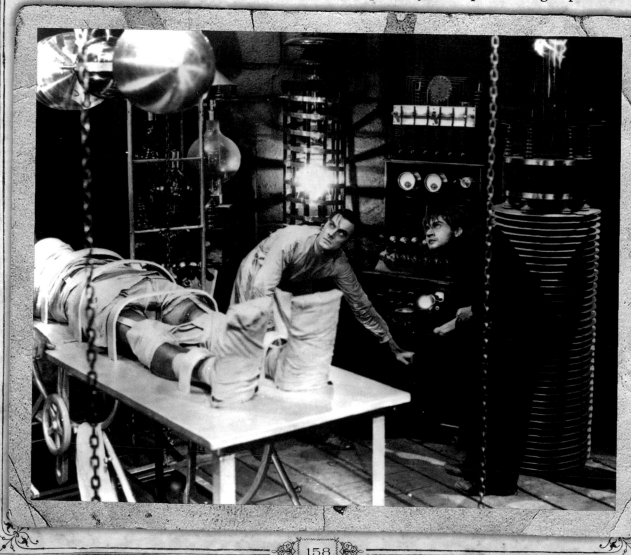

Dr Frankenstein (Colin Clive) utilizes a Tesla Coil to reanimate the Monster (Boris Karloff) helped by his assistant Fritz (Dwight Frye) in *Frankenstein* (1931).

wrote a paper denouncing 'Jewish physics' as the Nazis dubbed the new departures into relativity and quantum physics. Echoes of anti-Semitism can be found in Tesla's attacks on Einstein.

Returning to the US, deBobula designed a 120-ft (36-m) tower for Tesla's death beam. However, his involvement with a munitions company run by a German-American brought him to the attention of the IRS and the FBI. When his apartment was searched it was found to be full of grenades, tear-gas bombs and dynamite. Tesla was furious when deBobula used his name in an arms deal with Paraguay. Questioned by the authorities, deBobula denied any ties to the Communist Party or the German-American Bund which supported Hitler. The FBI monitored his activities throughout World War II. Nothing was ever proved against him. However, Tesla was tainted by association.[42]

REVIEWING HIS GREATEST INVENTIONS

The following year Tesla was still full of wild and abstruse pronouncements. He invited some 30 journalists to a gourmet luncheon to celebrate his 79th birthday in the private dining room of the Hotel New Yorker, where he was then staying. He had been thrown out of the Hotel Pennsylvania in 1930, owing $2,000, when other patrons complained of the pigeon droppings. While the reporters feasted at his expense, Tesla did not even touch a glass of water. However, towards the end of the meal, he went and got a small bottle of pasteurized milk which he poured into a silver dish and heated on a small oil stove beside the table. The Hotel New Yorker then supplied a birthday cake with one candle for their honoured guest.

Asked what was his greatest feat in the field of engineering, he said: 'An apparatus by which mechanical energy can be transmitted to any part of the terrestrial globe.'[43]

He called this discovery 'tele-geo-dynamics' and admitted that it would 'appear almost preposterous'. However, it would give the world a new means of unfailing communication, provide a new and by far the safest means for guiding ships at sea and into port, furnish a 'divining rod' for locating any type of ore beneath the surface of the Earth, and give scientists a means of 'laying bare the physical conditions of the Earth and enable them to determine all the Earth's physical constants'.[44]

The apparatus needed to do this, he said, was simple. It consisted of a stationary part and a cylinder of fine steel 'floating' in the air. He had, he said, discovered a means of 'impressing on the floating part powerful impulses which react on the stationary part, and through the latter to transmit energy through the Earth'. To do this he had 'found a new amplifier for a known type of energy'. The purpose was 'to produce impulses through the Earth and then pick them up whenever needed'.[45]

His second greatest invention, he said, 'will be considered absolutely impossible by any competent electrical engineer'. It was a new method of producing direct current without a commutator - something, he said, 'that has been considered impossible since the days of Faraday'.[46]

'Incredible as it seems,' he said, 'I have found a solution for this old problem.'

Next he came to cosmic rays which, he said, were produced by the force of electrostatic repulsion and consisted of powerfully charged positive particles that come to Earth from the Sun and other stars. 'After experimentation,' he said, 'the Sun is charged with an electrical potential of 215 billion volts, while the electric charge stored in the Sun amounted to around 50 billion billion electrostatic units.'[47]

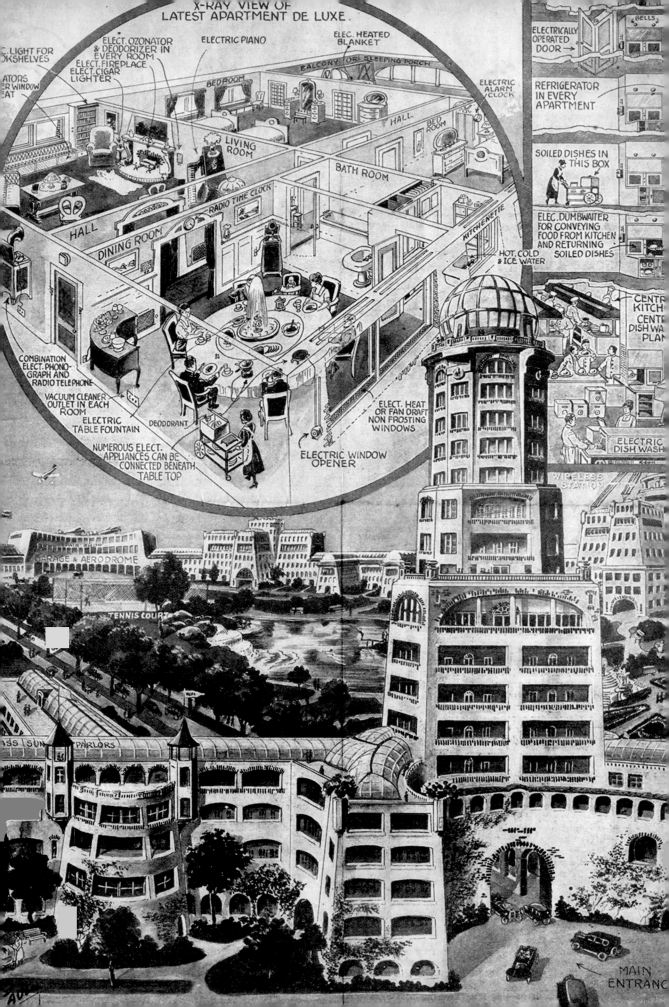

Again he dismissed the theory of relativity, describing it as 'a mass of error and deceptive ideas violently opposed to the teachings of great men of science and even to common sense'.

> *The theory, wraps all these errors and fallacies and clothes them in magnificent mathematical garb which fascinates, dazzles and makes people blind to the underlying errors. The theory is like a beggar clothed in purple whom ignorant people take for a king. Its exponents are brilliant men, but they are metaphysicists rather than scientists. Not a single one of the relativity propositions has been proved.*[48]

One of Tesla's great bugbears with relativity was its prohibition of anything travelling faster than the speed of light, which upset his theories about standing waves and the wireless transmission of energy. He was adamant that, in his observations of cosmic rays, he had already discovered particles that travelled faster than light.

> *In 1899, I obtained mathematical and experimental proofs that the Sun, and other heavenly bodies similarly conditioned, emit rays of great energy which consist of inconceivably small particles animated by velocities vastly exceeding that of light. So great is the penetrative power of these rays that they can traverse thousands of miles of solid matter with but slight diminution of velocity. In passing through space, which is filled with cosmic dust, they generated a secondary radiation of constant intensity, day and night, and pouring upon the earth equally from all directions. As the primary rays projected from the suns and stars can pass through distances measured in light-years without great diminution of velocity, it follows that whether a secondary ray is generated near a sun or at any distance from it, however great, its intensity is the same.*[49]

As of yet, no one else has found particles that travel faster than light. However, neutrinos generated by the fusion reactions in the Sun do have the penetrative power Tesla mentioned. They were predicted by the Italian-born physicist Enrico Fermi (1901 - 54) in 1934, but not detected experimentally for another 20 years. Nevertheless in 1932, Tesla said that he had 'harnessed the cosmic rays and caused them to operate a motive device'.[50] Cosmic rays, he said, struck the atmosphere, ionizing the air and creating charged particles - ions and electrons. 'These charges are captured in a condenser which is made to discharge through the circuit of a motor,' he said. He also said that he had 'hopes of building such a motor on a large scale'.[51] However, by 1935, he was also telling the *New York Herald Tribune* that some day the Sun would explode.[52]

THE TESLA INSTITUTE

In 1936, the Tesla Institute was opened in Belgrade, then the capital of Yugoslavia. A fully equipped research centre, it was funded by the Yugoslav government and private sources. A week of celebrations commemorating the great man's 80th birthday followed. They occurred in Belgrade on 26, 27 and 28 May, in Zagreb, capital of Croatia, on 30 May and in his native village of Smiljan on 2 June, and again on 12 July 1936.[53]

Left: An all-electric new De Luxe Apartment House, illustrated by Frank R. Paul, 1922.

CHAPTER 15

THE FINAL DAYS

What has the future in store for this strange being, born of a breath, of perishable tissue, yet Immortal, with his powers fearful and Divine? What magic will be wrought by him in the end? What is to be his greatest deed, his crowning achievement?

Nikola Tesla[1]

At 81, Tesla was honoured by both the Yugoslav and Czechoslovakian governments. He was awarded the Grand Cordon of the White Eagle, Yugoslavia's highest honour bestowed by King Peter through his regent Prince Paul, and Czechoslovakia's Order of the White Lion. These were presented at his birthday luncheon at the Hotel New Yorker. Meanwhile he was still making his birthday announcements of new discoveries to newspapermen.

In 1937, he told them that he had perfected the principle of a new tube that would make it possible to smash atoms and produce cheap radium. He would give a demonstration of it in 'only a little time'. He was expecting to put his discovery forward for France's Pierre Guzman Prize of 100,000 francs (around €50,000). It was awarded by the *Académie des Sciences* to 'the person of whatever nation who will find the means ... of communicating with a star and of receiving a response'.[2]

Tesla receives the Order of the White Lion from the Czechoslovakian government, 11 July 1937.

The money, of course, is a trifling consideration, but for the great historical honour of being the first to achieve this miracle I would be almost willing to give my life. I am just as sure that prize will be awarded to me as if I already had it in my pocket. They have got to do it. It means it will be possible to convey several thousand units of horsepower to other planets, regardless of distance. This discovery of mine will be remembered when everything else I have done is covered in dust.[3]

[The Guzman Prize was finally awarded to the crew of Apollo 11 in 1969 after the first Moon landing.]

His apparatus, Tesla said, employed more than three dozen of his own inventions. 'It is absolutely developed,' he said. 'I wouldn't be surer that I can transmit energy a hundred miles than I am of the fact that I can transmit energy a million miles up.'

It used a different kind of energy than was commonly employed which travels through a channel of 'less than one-half of one-millionth of a centimetre'. 'I could undertake a contract to manufacture the apparatus,' he said.

While he was certain that there was life on other planets, the problem with his equipment, he said, was hitting other moving planets with a needle-point of tremendous energy. But he thought that astronomers could help solve this problem. First they should aim this Tesla Ray at the Moon where they could easily see its effects - 'the splash and the volatilization of matter'. He also imagined advanced thinkers living on other planets were experimenting in this field, mistaking Tesla energy rays for cosmic rays.

On the practical front, he announced a new type of tube. His experiments had been rewarded with 'complete success' and he had 'produced a tube which it will be hard to improve further'.[4]

It is of ideal simplicity, not subject to wear and can be operated at any potential, however high – even 100 million volts – that can be produced. It will carry heavy currents, transform any amount of energy within practical limits and it permits easy control and regulation of the same. I expect that this invention, when it becomes known, will be universally adopted in preference to other forms of tubes and that it will be the means of obtaining results undreamed of before. Among others, it will enable the production of cheap

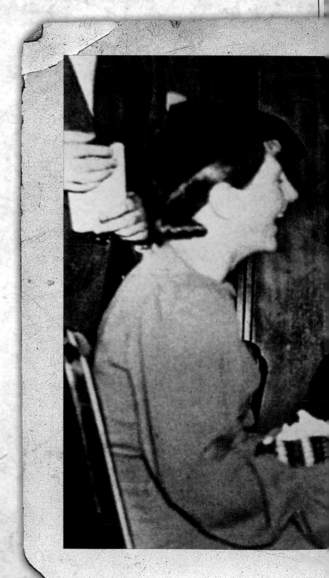

radium substitutes in any desired quantity and will be, in general, immediately more effective in the smashing of atoms and the transmutation of matter. However, this tube will not open up a way to utilize atomic or subatomic energy for power purposes. It will cheapen radium so that it will be just as cheap – well, it will get down to $1 a pound – in any quantity.[5]

Tesla was annoyed that some newspapers said he would be giving a full description of his invention at his birthday lunch. He could not release the information, he said, because he was bound by financial obligations involving 'vast sums of money'. And this was not an idle boast, he insisted: 'It is not an experiment. I have built, demonstrated and used it. Only a little time will pass before I can give it to the world.'[6]

The New York Times, perhaps with tongue in cheek, went on to report that Dr Tesla entertained his guests with colourful personal reminiscences and observations including his opinions on dieting and immortality.

Tesla being interviewed by reporters, 10 January 1935.

MORE MONEY WORRIES

Although Tesla's mind was as fertile as ever, his financial situation continued to decline. When Hugo Gernsback showed him Westinghouse's latest radio set, Tesla saw immediately that they were flagrantly infringing his wireless patents. He protested, but was in no position to fight a large corporation.

Unable to pay his hotel bill, again, Tesla handed over the 'working model' of his death beam as collateral. It was worth, he said, $10,000. He also told Jack Morgan that the Russians were keen to buy his death beam to defend themselves against the Japanese. However, he already owed Morgan a great deal of money over his bladeless turbines and, despite filling his letters with attacks on Franklin Roosevelt's 'New Deal' with Astor who Morgan hated, no money was forthcoming.

Eventually, Westinghouse acknowledged Tesla's contribution to the company and paid him $125 a month as a consulting engineer. They also came to an agreement with the Hotel New Yorker where Tesla lived rent free for the rest of his life. In his last years, the Yugoslav government also gave him an honorarium of $7,200 a year. This allowed him to give generous tips to those who had rendered him the slightest assistance and hand-outs, that he could ill-afford, to anyone he thought was in need.[7]

TESLA'S LAST INTERVIEWS

Tesla gave some of his last interviews to Nazi apologist George S. Viereck. Again he explained that he was not a believer in God in the conventional sense. Perhaps under Viereck's influence, Tesla espoused eugenics - the forced sterilization of those thought to be mentally unfit - which were then being practised in the US as well as Nazi Germany.

The year 2100 will see eugenics universally established. In past ages, the law governing the survival of the fittest roughly weeded out the less desirable strains. Then man's new sense of pity began to interfere with the ruthless workings of nature. As a result, we continue to keep alive and to breed the unfit. The only method compatible with our notions of civilization and the race is to prevent the breeding of the unfit by sterilization and the deliberate guidance of the mating instinct. Several European countries and a number of states of the American Union sterilize the criminal and the insane. This is not sufficient. The trend of opinion among eugenists is that we must make marriage more difficult. Certainly no one who is not a desirable parent should be permitted to produce progeny. A century from now it will no more occur to a normal person to mate with a person eugenically unfit than to marry a habitual criminal.[8]

Although at the time, the environment was hardly on the agenda, Tesla told Viereck:

Hygiene, physical culture will be recognized branches of education and government. The Secretary of Hygiene or Physical Culture will be far more important in the cabinet of the President of the United States who holds office in the year 2035 than the Secretary of War. The pollution of our beaches such as exists today around New York City will seem as unthinkable to our children and grandchildren as life without plumbing seems to us. Our water supply will be far more carefully supervised, and only a lunatic will drink unsterilized water.[9]

He looked forward to a time where science and education would be more important than war:

Today the most civilized countries of the world spend a maximum of their income on war and a minimum on education. The 21st century will reverse this order. It will be more glorious to fight against ignorance than to die on the field of battle. The discovery of a new scientific truth will be more important than the squabbles of diplomats. Even the newspapers of our own day are beginning to treat scientific discoveries and the creation of fresh philosophical concepts as news. The newspapers of the 21st century will give a mere 'stick' in the back pages to accounts of crime or political controversies, but will headline on the front pages the proclamation of a new scientific hypothesis.[10]

Now gaunt from his meagre diet, he had clear views on the future of food:

More people die or grow sick from polluted water than from coffee, tea, tobacco, and other stimulants. I myself eschew all stimulants. I also practically abstain from meat. I am convinced that within a century, coffee, tea, and tobacco will no longer be in vogue. Alcohol, however, will still be used. It is not a stimulant but a veritable elixir of life. The abolition of stimulants will not come about forcibly. It will simply be no longer fashionable to poison the system with harmful ingredients. [Bodybuilder] Bernarr Macfadden has shown how it is possible to provide palatable food based upon natural products such as milk, honey, and wheat. I believe that the food which is served today in his penny restaurants will be the basis of epicurean meals in the smartest banquet halls of the 21st century.

There will be enough wheat and wheat products to feed the entire world, including the teeming millions of China and India, now chronically on the verge of starvation. The earth is bountiful, and where her bounty fails, nitrogen drawn from the air will refertilize her womb. I developed a process for this purpose in 1900. It was perfected 14 years later under the stress of war by German chemists.[11]

After subsisting on a diet of bread, warm milk and what he called 'Factor Actus', Tesla gave up solid food altogether, living on thin gruel of cauliflower, leeks, cabbage, turnips and lettuce. But he was still strong enough to make predictions. Then, in his last days, he lived on milk and honey, believing them to be the purest foods. Nevertheless, the future looked bright:

Long before the next century dawns, systematic reforestation and the scientific management of natural resources will have made an end of all devastating droughts, forest fires, and floods. The universal utilization of water power and its long-distance transmission will supply every household with cheap power and will dispense with the necessity of burning fuel. The struggle for existence being lessened, there should be development along ideal rather than material lines.[12]

What's more, the work would be done by robots, something Tesla had been working on for nearly 40 years.

At present we suffer from the derangement of our civilization because we have not yet completely adjusted ourselves to the machine age. The solution of our problems does not lie in destroying but in mastering the machine.

Innumerable activities still performed by human hands today will be performed by automatons. At this very moment scientists working in the laboratories of American universities are attempting to create what has been described as a 'thinking machine'. I anticipated this development.

I actually constructed 'robots'. Today the robot is an accepted fact, but the principle has not been pushed far enough. In the 21st century the robot will take the place which slave labour occupied in ancient civilization. There is no reason at all why most of this should not come to pass in less than a century, freeing mankind to pursue its higher aspirations.[13]

GEORGE SYLVESTER VIERECK (1884 – 1962)

One of Tesla's more unusual friends towards the end of his life was the poet and author George Sylvester Viereck. Born in Munich, he was thought to have been the illegitimate grandson of Kaiser Wilhelm I, an idea he encouraged. He emigrated to the US with his family in 1896 and published his first book of poetry *Gedichte* in 1904 while still a student at the College of the City of New York. It received some favourable critical attention and he quickly built a reputation. But then came World War I. He said: 'In spite of my staunch support of the American war effort, I was decried as an isolationist and a pro-German. I was boycotted by the war party. Five celebrated authors banded themselves together under the slogan, *Never Again Viereck*. My verse was dropped from anthologies and my name from *Who's Who* in America. I was expelled from the Poetry Society of America which owed its existence largely to my efforts and from the Author's League. I was now a poet without a licence.'[14]

He then laboured to build a reputation as a novelist and, as a journalist, interviewed Sigmund Freud, George Bernard Shaw, Mussolini, Hindenburg, Grand Duke Alexander of Russia, Einstein, Henry Ford and, of course, Tesla. With the advent of World War II, he was again thought to be pro-German. This time with good reason. He was found to have accepted money from the Nazi Party to spread propaganda in the US and was jailed. After the war, his memoir of his prison experiences *Men Into Beast* sold well, but he never regained his literary reputation. Tesla knew all his poems by heart.

ELMER GERTZ (1906 – 2000)

Another of the dinner guests Tesla met at Viereck's house on Riverside Drive was the attorney Elmer Gertz. He was a friend of the poet Carl Sandburg (1878 – 1967) and the biographer and agent of British journalist and voluptuary Frank Harris (1856 – 1931). He also defended the work of American writer Henry Miller (1891 – 1980) in obscenity trials, murderer Nathan Leopold (1904 – 71) in the 1924 Leopold and Loeb case and club owner Jack Ruby (1911 – 67) who murdered President Kennedy's assassin Lee Harvey Oswald (1939 – 63). He was impressed that he had seen Tesla in the same house as novelist Sinclair Lewis (1885 – 1951), Einstein and countless others. He said: 'In a communicative mood, Tesla told his life story unostentatiously, simply, with quiet eloquence. He told of his platonic affairs of the heart ... explained the inventions that have made the world his debtor ... and told of his plans, of his credo, of his foibles. It was a tale of wonder, told with gulleless simplicity.'[15]

WORLD WAR II

Tesla was growing feeble, but with the help of his nephew Sava Kosanovic he wrote a foreword to a Serbo-Croat edition of *The Future of the Common Man* by the then Vice-President Henry Wallace (1888 - 1965). In it, he said: 'Out of this war, the greatest since the beginning of history, a new world must be born that would justify the sacrifices offered by humanity, where there will be no humiliation of the poor by the violence of the rich; where the products of intellect, science and art will serve society for the betterment and beautification of life, and not the individuals for achieving wealth. This new world shall be a world of free men and free nations, equal in dignity and respect.'[16]

While the regent of Yugoslavia, Prince Paul (1893 - 1976), sought to make a treaty with the Nazis, King Peter (1923 - 70) opposed it. When he came of age at 17, Germany invaded and the king went into exile. Sava Kosanovic went with him. However, he began to favour the Communist guerrilla leader Josip Broz Tito (1892 - 1980), who the British and Americans were also reluctantly backing. However, in the US, Kosanovic brought King Peter to Tesla's hotel. In his diary King Peter wrote:

I visited Dr Nikola Tesla, the world-famous Yugoslav-American scientist, in his apartment in the Hotel New Yorker. After I had greeted

him the aged scientist said: 'It is my greatest honour. I am glad you are in your youth, and I am content that you will be a great ruler. I believe I will live until you come back to a free Yugoslavia. From your father you have received his last words: "Guard Yugoslavia." I am proud to be a Serbian and a Yugoslav. Our people cannot perish. Preserve the unity of all Yugoslavs – the Serbs, the Croats and the Slovenes.'[17]

The two of them wept over the fate of their homeland.

GENEROUS TO THE END

Although Tesla was dogged by his own financial problems, he was generous to the end. A few days before he died, he called one of his favourite messenger boys, a lad named Kerrigan, and gave him an envelope, addressed to: *Mr Samuel Clemens, 35 South Fifth Avenue, New York City*. He told him it was to be delivered as quickly as possible. After a while the boy returned, saying that there was no such street as South Fifth Avenue. Tesla was furious. Mr Clemens was a very famous author who wrote under the name of Mark Twain, he told Kerrigan, and the address he had given was correct.

Kerrigan tried again. And when he had no luck he reported to his office manager, who quickly spotted the boy's difficulty. South Fifth Avenue had changed its name to West Broadway years before and Mark Twain had been dead over 30 years. The address Tesla had given him was that of his laboratory that had burnt down. Kerrigan returned to Tesla to explain his difficulties. Tesla was outraged when the boy told him that Mark Twain was dead. 'He was in my room here last night,' Tesla insisted. 'He is having financial difficulties and needs my help.' And he sent the boy to deliver the envelope again. Confused Kerrigan returned to his manager who opened the envelope to see whether it contained any clue to where it should be delivered. All that was inside was a blank piece of paper and $100 in $5 bills. When Kerrigan returned the envelope to Tesla yet again, the inventor told him that, if he could not deliver the money, he should keep it.[18]

At the same time, while Tesla could not pay the $297 for his possessions that were being kept in storage, he sent a cheque for $500 to a Serbian Church in Gary, Indiana. Tesla's biographer John J. O'Neill saw an advertisement for Tesla's possessions that were being sold off by the storage company to cover the bill and contacted the inventor's nephew Sava Kosanovic who paid the bill, preventing the loss of Tesla's priceless papers.[19]

THE SLIGHT HINT OF A SMILE

During the latter part of 1942, Tesla became practically a recluse. Physically weak, he retired to bed and permitted no visitors. Hotel staff were not to visit his room unless he summoned them and he refused to listen to any suggestion that they call a doctor. On 5 January 1943, he called a maid to his room and issued orders that he was not to be disturbed. Nothing was heard from him for three days. Finally the maid decided to risk his wrath and check up on him. She entered the room in trepidation and found him dead, with the slight hint of a smile on his gaunt face.

As he had died alone without medical attention, the police were called. The Medical Examiner put the time of death as 10.30 pm on Thursday 7 January, just a few hours before the maid's early morning visit. Tesla had died in his sleep. The cause of death was given as coronary thrombosis and the Medical Examiner noted that there were 'No suspicious circumstances.'[20]

Agents from the FBI came to open the safe in Tesla's room and read his papers

FRAGMENTS OF OLYMPIAN GOSSIP

Under the influence of Viereck, Tesla, who was always competitive, wrote a poem call *Fragments of Olympian Gossip*, poking fun at the scientific establishment:

While listening on my cosmic phone
I caught words from the Olympus blown.
A newcomer was shown around;
That much I could guess, aided by sound.
There's Archimedes with his lever
Still busy on problems as ever.
Says: matter and force are transmutable
And wrong the laws you thought immutable.
Below, on Earth, they work at full blast
And news are coming in thick and fast.
The latest tells of a cosmic gun.
To be pelted is very poor fun.
We are wary with so much at stake,
Those beggars are a pest – no mistake.
Too bad, Sir Isaac, they dimmed your renown
And turned your great science upside down.
Now a long haired crank, Einstein by name,
Puts on your high teaching all the blame.
Says: matter and force are transmutable
And wrong the laws you thought immutable.
I am much too ignorant, my son,
For grasping schemes so finely spun.
My followers are of stronger mind
And I am content to stay behind,
Perhaps I failed, but I did my best,
These masters of mine may do the rest.
Come, Kelvin, I have finished my cup.
When is your friend Tesla coming up.
Oh, quoth Kelvin, he is always late,
It would be useless to remonstrate.
Then silence – shuffle of soft slippered feet –
I knock and – the bedlam of the street.[21]

EDWIN ARMSTRONG (1890 – 1954)

At the age of 14, Armstrong read of the achievements of Marconi and set about building wireless equipment in the attic of his family home in New York. He studied under Pupin at Columbia, where he developed the regenerative, or feedback, circuit using De Forest's Audion tube. This increased the amplification a thousand-fold. Signals that could barely be distinguished through headphones could now be heard across the room. At its highest amplification, the circuit became an oscillator that generated radio waves and became the basis of radio and television broadcasting. However, Armstrong was challenged by De Forest in a series of patent suits which dragged on for 14 years.

In the army during World War I, Armstrong improved Fessenden's heterodyne circuit to make the superheterodyne circuit, which again greatly improved the sensitivity and stability of radio signals that underlies nearly all radio, radar and television reception over the airwaves. He sold the patents to Westinghouse, which made him, briefly, a millionaire, and he married Marion MacInnis, secretary to David Sarnoff, then general manager of RCA.

Armstrong went on to invent FM, or frequency modulation. By modulating the frequency of the signal rather than its amplitude, or strength, as in AM, high-fidelity broadcasting became possible. But the well established AM broadcasters were resistant. Armstrong had to spend $300,000 on building an FM station to prove its worth. After World War II, Armstrong began a patent-infringement suit against RCA who had adopted FM for television. Impoverished by litigation, Armstrong jumped to his death from a 13th storey window.

in case there was anything in them that might aid the war effort. However, Hugo Gernsback, Kenneth Swezey, Sava Kosanovic and George Clark of RCA had already entered the apartment. While Gernsback went to organize a death mask, the other three had the safe opened by a locksmith with representatives of the hotel management present. The FBI said that valuable papers were taken. However, the hotel management said that Kosanovic only took three pictures and Swezey took the testimonial book created for Tesla's 75th birthday. However, Kosanovic was sure that someone had already gone through his uncle's effects. Technical papers were missing, along with a black notebook he knew that Tesla kept. It contained several hundred pages of notes, some of which were marked 'government'.[22]

STATE FUNERAL

The Yugoslav ambassador Dr Constantin Fotitch laid on a state funeral for Tesla at the Cathedral of St John the Divine. Over 2,000 mourners were present, including other inventors. While the church was Episcopalian, the service was Orthodox and conducted in Serbian. The honorary pallbearers included Dr Ernst Alexanderson of General Electric who patented a high-frequency transmitter, Edwin Armstrong, father of FM radio, Gano Dunn, president of J.G. White Engineering who had been Tesla's assistant at his ground-breaking lecture at Columbia, and representatives from Westinghouse, Columbia University and the Hayden Planetarium where Tesla would often go to meditate. A number of Yugoslav ministers were there. King Peter II of Yugoslavia sent a wreath and the chief mourner was Tesla's nephew Sava Kosanovic, who was by then president of the Eastern and Central European

Planning Board, representing Yugoslavia, Czechoslovakia, Poland and Greece. Later he became Yugoslav ambassador to the US.[23]

Scientists paid tribute to his intellect and technological achievements, and telegrams of condolence came from Nobel Prize winners, prominent scientists, literary figures and US officials. A message from Mrs Eleanor Roosevelt read: 'The President and I are deeply sorry to hear of the death of Mr Nikola Tesla. We are grateful for his contribution to science and industry and to this country.'[24]

Vice-President Henry Wallace paid a more personal tribute: 'Nikola Tesla, Yugoslav born, so lived his life as to make it an outstanding example of that power that makes the United States not merely an English-speaking nation but a nation with universal appeal. In Nikola Tesla's death the common man loses one of his best friends.'[25]

Over the radio New York Mayor Fiorello La Guardia (1882–1947) read a eulogy written by Slovene author Louis Adamic.[26] After the service, Tesla's body was taken to Ferncliff Cemetery in Ardsley, New York, and was later cremated.[27]

TRIBUTES TO TESLA

Tesla had always been loved by the popular press for his shocking experiments and outrageous pronouncements. On his death, the *New York Sun* wrote:

Mr Tesla was 86 years old when he died. He died alone. He was an eccentric, whatever that means. A non-conformist, possibly. At any rate, he would leave his experiments and go for a time to feed the silly and inconsequential pigeons in Herald Square. He delighted in talking nonsense; or was it? Granting that he was a difficult man to deal with, and that sometimes his predictions would affront the ordinary human's intelligence, here, still, was an extraordinary man of genius. He

Left: Edwin Armstrong and his new invention, the super-regenerative receiver in 1922.

must have been. *He was seeing a glimpse into that confused and mysterious frontier which divides the known and the unknown … But today we do know that Tesla, the ostensibly foolish old gentleman at times was trying with superb intelligence to find the answers. Probably we shall appreciate him better a few million years from now.*[28]

More tributes came rolling in. Hugo Gernsback wrote: 'We cannot know, but it may be that a long time from now, when patterns have changed, the critics will take a view of history. They will bracket Tesla with Da Vinci, or with our own Mr Franklin … One thing is sure. The world, as we run it today, did not appreciate his peculiar greatness.'[29]

The President of RCA David Sarnoff said: 'Nikola Tesla's achievements in electrical science are monuments that symbolize America as a land of freedom and opportunity … His novel ideas of getting the ether in vibration put him on the frontier of wireless. Tesla's mind was a human dynamo that whirled to benefit mankind.'[30]

Edwin Armstrong, who went on to sue RCA for infringing his patents, said:

Who today can read a copy of The Inventions, Researches and Writings of Nikola Tesla, published before the turn-of-the-century, without being fascinated by the beauty of the experiments described and struck with admiration for Tesla's extraordinary insight into the nature of the phenomena with which he was dealing? Who now can realize the difficulties he must have had to overcome in those early days? But one can imagine the inspirational effect of the book 40 years ago on a boy about to decide to study the electrical art. Its effect was both profound and decisive.[31]

Nine months after Tesla's death the USS *Nikola Tesla* - a Liberty ship, vital to the Allied war effort - was launched in Baltimore.[32]

THE MISSING PAPERS

The papers from the safe in Tesla's room were lodged with the Office of Alien Property. This was unusual as Tesla was a US citizen. The day after Tesla died Abraham N. Spanel, the designer of floating pontoon stretchers, got in touch with FBI agent Fredrich Cornels about the Tesla papers, fearing that, via Kosanovic, they might fall into Communist hands. Before he had started the International Latex Company of Dover, Delaware (who later made space suits), Spanel had worked as an electrical engineer and considered that some of Tesla's unmade inventions might play a vital role in the war. He made waves in high places, contacting others in the White House and the FBI.

Spanel also got in touch with the young electrical engineer Bloyce Fitzgerald who had also contacted Cornels. Fitzgerald had been writing to Tesla for some time. Finally he got to meet Tesla just a few weeks before his death. At the time, Fitzgerald was at the Massachusetts Institute of Technology working on the 'dissipation of energy from rapid-fire weapons'. Around the same time Tesla had complained that someone had been trying to steal his invention. His room had been entered and his papers examined, but 'the spies had left empty handed,' he said.

Fitzgerald told Cornels that, among Tesla's papers, were complete plans for his death beam, along with specifications and explanations of how the thing worked. There was also 'a working model … which cost more than $10,000 to build in a safety deposit box of Tesla's at the Governor Clinton Hotel'.

Charles Hausler, one of Tesla's pigeon handlers, also said that Tesla kept a large box in his room near his pigeon cages. 'He told me to be very careful not to disturb the box as it contained something that could destroy an airplane in the sky and he had hopes of presenting it to the world,' Hausler said.

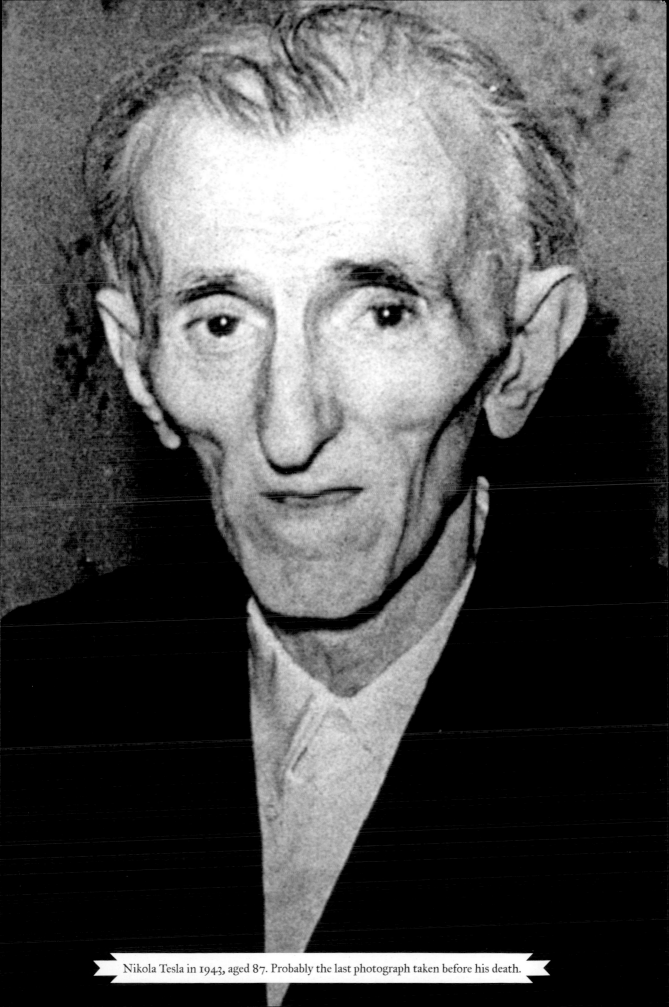

Nikola Tesla in 1943, aged 87. Probably the last photograph taken before his death.

TRUNKS FULL OF ALIEN PROPERTY

According to Fitzgerald, Tesla had claimed that he had some 80 trunks in various places around the city. It was vital for the war effort that the government get their hands on the Tesla papers. He also expressed doubts about the loyalty of nephews Sava Kosanovic and Nicholas Trbojevich. There was even talk of having Kosanovic and Swezey arrested for theft, but as the papers were now with the Office of Alien Property, this seemed unnecessary.

Cornels' boss P.E. Foxworth, assistant director of the New York FBI office, was called in to investigate. The government, he said, was 'vitally interested' in preserving Tesla's papers.

The legality of the OAP holding onto Tesla's papers was questionable. Although legal title had fallen to Kosanovic, who was a foreigner, the government agency was not permitted to seize 'enemy assets' without

a court order. Nevertheless Alien Property custodian Irving Jurow got a call ordering him to impound all of Tesla's belongings as he was reputed to have invented a death ray that disintegrated enemy planes in the sky and German agents were after it. Along with agents from Naval Intelligence, Army Intelligence and the FBI, Jurow visited the Hotel New Yorker, the Governor Clinton, the St Regis, the Waldorf-Astoria and the storage facility where Tesla had kept the rest of his things. Papers in possession of John J. O'Neill, who was preparing a biography of Tesla, were also confiscated. These were perused by various branches of the military.

Dr John G. Trump, an electrical engineer with the National Defense Research Committee of the Office of Scientific Research and Development, was also called to analyze the Tesla papers in OAP custody. Following a three-day investigation, Dr Trump concluded that Tesla's 'thoughts and efforts during at least the past 15 years were

Tesla meets King Peter II of Yugoslavia on 15 July 1942. Tesla's nephew, Sava Kosanovic, is third from the left.

primarily of a speculative, philosophical, and somewhat promotional character often concerned with the production and wireless transmission of power; but did not include new, sound, workable principles or methods for realizing such results'. However, in his report, he admits that he had not bothered to open many of the trunks, so confident was he that nothing of value would be found.[33]

THE PARTICLE BEAM

However, among the papers was the still unpublished 'The New Art of Projecting Concentrated Non-Dispersive Energy Through Natural Media'. This was classified top secret and distributed to Naval Intelligence, the National Defense Research Council, the FBI, MIT, Wright-Patterson Air Force Base - where the V1 flying bomb was reverse engineered - and, probably, the White House.

The paper described in detail the new features, including how to create a non-dispersive beam of particles:

I perfected means for increasing enormously the intensity of the effects, but was baffled in all my efforts to materially reduce dispersion and became fully convinced that this handicap could only be overcome by conveying the power through the medium of small particles projected, at prodigious velocity, from the transmitter. Electro-static repulsion was the only means to this end and apparatus of stupendous force would have to be developed, but granted that sufficient speed and energy could be realized with a single row of minute bodies then there would be no dispersion whatever even at great distance. Since the cross section of the carriers might be reduced to almost microscopic dimensions an immense concentration of energy, irrespective of distance, could be attained.[34]

Then there was a diagram of the open-ended vacuum where, he said:

It will be observed that in this tube I do away with the solid wall or window indispensable in all types heretofore employed, producing the high vacuum required and preventing the inrush of the air by a gaseous jet of high velocity. Evidently, to secure this result, the dynamic pressure of the jet must be at least equal to the external static pressure.

The high-potential generator, creating a charge of up to 100 million volts, worked on the principle of a Van de Graaf generator - with the belt replaced by a high-speed stream of ionized air. The charges would be stored in specially shaped bulbs around the top of a 100 ft (30 m) tower. In the super-gun itself, tungsten wire would be fed into the high-vacuum firing chamber. Under the huge electrostatic forces generated, tiny droplets of metal would be sheared off and projected out of the chamber at over 400,000 feet per second (120 km per second).

John Trump also went to the Hotel Governor Clinton to examine the working model that Tesla had said he left there. Tesla had told the management that the box containing a secret weapon that was rigged to detonate if it was opened by an unauthorized person. Trump approached with trepidation. All he found inside was 'a multi-decade resistance box of the type used for Wheatstone bridge resistance measurements - a common standard item to be found in every electrical laboratory before the turn of the century'.[35] Trump took no further interest.

THE VANISHING DEATH RAY

In 1937, Tesla had claimed that he had built his death beam, saying: 'It is not an experiment. I have built, demonstrated and used it. Only a little time will pass before I can give it to the world.'[36] At the time, he did have two secret labs where journalists were not allowed.

Mrs Czito, whose husband's father and

grandfather both worked for Tesla, said that her father-in-law said Tesla had a device that could bounce a beam off the Moon. There is also a story that the deBobula design was taken to Alcoa Aluminium who said they would furnish the material required once the capital had been raised.

There is, of course, a conspiracy theory to account for the missing model. When the safe in Tesla's room was opened, the locksmith was asked to change the combination. Inside the safe, when it was locked again, was a set of keys and Tesla's 1917 Edison Medal. The new combination was given to Kosanovic and no one else. When the safe was finally shipped to Belgrade and opened some 10 years later, the medal and the keys were missing. The medal has never been found, but the keys were found in one of the numerous cases of documents that accompanied the safe in the shipment.

According to the Office of Alien Property, the two representatives of the hotel management who were present were Mr L.O. Doty, credit manager, and Mr L.A. Fitzgerald, assistant credit manager. We have already met Bloyce Fitzgerald in connection with Tesla, but Tesla's biographer Marc Seifer has also unearthed a letter from a Colonel Ralph E. Doty, chief of the Washington Branch of the Military Intelligence Service. Dated 22 January 1946, it is addressed to the Alien Property Custodian and reads:

Dear Sir: This office is in receipt of a communication from Headquarters, Air Technical Service Command, Wright Field, requesting that we ascertain the whereabouts of the files of the late scientist, Dr Nikola Tesla, which may contain data of great value to the above Headquarters. It has been indicated that your office might have these files in custody. If this is true, we would like to request your consent for a representative of the Air Technical Service Command to review them.

In view of the extreme importance of these files to the above command, we would like to request that we be advised of any attempt by any other agency to obtain them. Because of the urgency of this matter, this communication will be delivered to you by a Liaison Officer of this office in the hope of expediting the solicited information.[37]

If the Fitzgerald and Doty at the opening of the safe were government agents, they might easily have got hold of the combination and the keys. It seems that Spanel was already under surveillance by the FBI for his pro-Communist sympathies. In November 1942, he met Fitzgerald at an engineering meeting, shortly before Tesla's death. At the time Fitzgerald was an army private at Wright Field. The FBI report later describes him as 'a brilliant 20-year-old scientist who spent hours with Tesla before his death ... Fitzgerald had developed some sort of anti-tank gun'.[38] Spanel tried to form a partnership with Fitzgerald to sell the weapon to the Remington Arms Company, but the deal fell through when a more lucrative offer came from the Eiogens Ship Building Company.

Fitzgerald was fired by Eiogens in November 1943 and returned to Wright Field as a private. According to his FBI file, in 1945, he was 'engaged in a highly secret experimental project at Wright Field ... In spite of his rank as private, Fitzgerald is actually director of this research and is working with many top young scientists ... on the perfection of Tesla's 'death ray' which in Fitzgerald's opinion is the only defence against the offensive use by another nation of the atomic bomb.'[39]

The files that Colonel Doty requested did find their way to Wright Field because, on 24 October 1947, the Director of the Office of Alien Property wrote to the commanding officer of the Air Technical Service Command, asking for them back.

The following month, a Colonel Duffy wrote back, saying: 'These reports are now in the possession of the Electronic Subdivision and are being evaluated. This should be completed by January 1, 1948. At that time your office will be contacted with respect to final disposition of these papers.'[40]

BELGRADE BOUND

In 1946, Sava Kosanovic had become Yugoslavian ambassador to the US after the US recognized Tito's Communist government. But by 1950, he was still not allowed access to Tesla's effects. He made official representations and in 1952, 80 trunks containing Tesla's papers, equipment and other belongings were shipped to Belgrade.[41] Five years later, Tesla's ashes followed. In the Tesla Museum they were placed in a spherical urn as this was Tesla's favourite shape.

This did not stop *American Scientist* being interested in Tesla's papers. Regularly individuals contacted the FBI who were thought to have made microfilm copies of everything sent to Yugoslavia. In the 1970s, it was thought that Tesla's fireballs might be a way of containing nuclear fusion. During the Cold War, both sides again investigated the possibility of making a particle beam to disable incoming nuclear missiles. Naturally Tesla's work would have been a starting point.

In 1977, *Aviation Week & Space Technology* published an article called *Soviets Push for Beam Weapon*.[42] In it, the retired head of Air Force Intelligence, General George J. Keegan said that the Soviet Union was attempting to develop a charge-particle beam at the test facility near the city of Semipalatinsk in Kazakhstan. The editor of *Aviation Week*, Robert Holtz said:

The Soviet Union has achieved a technical breakthrough in high-energy physics applications that may soon provide it with a directed-energy beam weapon capable of neutralizing the entire United States ballistic missile force and checkmating this country's strategic doctrine ... The race to perfect directed-energy weapons is a reality. Despite initial skepticism, the US scientific community is now pressuring for accelerated efforts in this area.[43]

The news got through to the White House, but President Jimmy Carter (1924-) said: 'We do not see any likelihood at all, based on our constant monitoring of the Soviet Union, that they have any prospective breakthrough in a new weapons system that would endanger our country.'[44]

Nevertheless, under the Carter administration, work began on a major space-based laser programme. Under the direction of the Defense Advanced Research Projects Agency, the ALPHA chemical laser project was begun in 1978. Contracts for the TALON GOLD targeting system were awarded in 1979, and the Large Optics Demonstration Experiment (LODE) started in 1980. These programmes formed the basis for the Strategic Defense Initiative - aka 'Star Wars' - announced by President Ronald Reagan (1911 - 2004) in 1983. SDI added X-ray lasers and neutral particle beams, all of which were reminiscent of Tesla's 'Chinese wall'. While this work was going on, fresh applications were made to the FBI to release any copies of the Tesla files they had, if only to know what the Soviets may had learned about beam weapons in Belgrade.[45]

However, after the collapse of the Soviet Union it was discovered that the test site outside Semipalatinsk was not working on a beam weapon at all, but rather on nuclear-powered rockets. John Pike, defence analyst at the Federation of American Scientists, said that the misidentification of the Semipalatinsk test site 'must rank as one of the major intelligence failures of the Cold War'.[46]

CHAPTER 16

POSTHUMOUS RECOGNITION

I misunderstood Tesla. I think we all misunderstood Tesla. We thought he was a dreamer and visionary. He did dream and his dreams came true, he did have visions but they were of a real future, not an imaginary one.

John Stone Stone[1]

In 1943 - a few months after Tesla's death - the US Supreme Court upheld Tesla's radio patent number 645,576. The Court had a selfish reason for doing so. The Marconi Company was suing the United States Government for use of its patents in World War I. The Court simply avoided the action by restoring the priority of Tesla's patent over Marconi. Nevertheless, it re-established the situation that had applied 40 years earlier when Tesla was recognized as the inventor of radio whose patents had been infringed by Marconi.

In 1956, the Nobel laureate Niels Bohr (1885 - 1962), father of quantum physics, spoke at a centennial congress held in Tesla's honour. In Yugoslavia, a commemorative stamp was issued. Tesla appeared on the 100 dinar note and the 'Tesla' became the unit of magnetic flux density. The asteroid 2244 *Tesla* was named by Serbian astronomer Milorad Protić in Belgrade when he discovered it on 22 October 1952. The Tesla Crater on the far side of the Moon was named after him when five Lunar Orbiters photographed the other side in 1966 and 1967. The IEEE has presented the Nikola Tesla Award for outstanding contributions to the generation or utilization of electric power annually since 1976. Nikola Tesla Corner is located in the heart of Manhattan, at the corner of West 40th Street and 6th Avenue reminding New Yorkers that the great man used to roam those very streets. Niagara Falls also celebrates the legendary inventor with two monuments, one on Goat Island was unveiled in 1976, while in 2006 a statue of Tesla standing on an AC motor was erected in Queen Victoria Park on the Canadian side of the falls. Finally, back in Serbia, Belgrade International Airport was renamed Belgrade Nikola Tesla Airport in 2006.

During the second half of the 20th century Tesla was overlooked by the history books. Unlike Marconi and Edison, he did not have large companies named after him. Westinghouse could have honoured him, but not unnaturally they give their plaudits to George Westinghouse himself. Also unlike Westinghouse, Edison and the Wright Brothers, Tesla was not born in America.[2] Those great Americans were seen as practical men who made things that ordinary people could understand. Few people in the general populace understand what AC is, how an induction motor works or what the principles behind the propagation of radio waves are.

With Tesla's seemingly science-fiction ideas towards the end of his life, he was seen as a mystic and, more than anything, the ultimate mad scientist with his middle-European accent and his eccentric personal habits. However, in the 1950s, Tesla's fame grew in the burgeoning counter-culture. An electrical engineer name Arthur H. Matthews claimed to have made a Teslascope to communicate with the inhabitants of other planets. He said that Tesla was a Venusian born on a spaceship travelling from Venus to Earth in July 1856, revealing all in *Wall of Light: Nikola Tesla and the Venusian Spaceship*. In 1959, Margaret Storm published the occult biography of Tesla, *Return of the Dove*. Her story also involved flying saucers which were very much in vogue.

In the 1970s, Ralph Bergstresser, who had known Tesla for the last 6 months of his life, produced 'Tesla Plates' which are marketed by the Swiss Tesla Institute. These, in some fashion, are supposed to tap into the energy of the universe. As Tesla's ideas, especially in later life, drew on thinking from both the East and the West, he became of interest to New Age thinkers. His striking, dark, middle-European looks and his development of willpower also appealed.

With the energy crisis of the 1970s, Tesla's idea of free energy had renewed

PROPHET OF SCIENCE

NIKOLA TESLA IS THE FORGOTTEN GENIUS, WHO REVOLUTIONIZED THE SCIENCE OF ELECTRICITY AND MADE FANTASTIC PREDICTIONS WHICH TIME PROVED TRUE.

NIKOLA WAS BORN IN 1857 AT SMILGAN, CROATIA. ONE DAY...

NOW OUR NEW FIRE ENGINE WILL START TO PUMP WATER... BUT SOMETHING'S WRONG.

I'LL FIX IT.

THERE GOES NIKOLA. WHAT DOES HE KNOW?

BUT YOUNG NIKOLA KNEW THE HOSE WAS CLOGGED AND

HE'S DONE IT! WATCH OUT FOR THE WATER!

appeal and a researcher named Robert Golka tried to recreate Tesla's Colorado experiments. In 1984, the first International Tesla Symposium was held in Colorado Springs. The group met annually for the next 14 years. It spawned the International Tesla Society whose membership grew to 7,000 worldwide, but it went bust in 1998. Devotees have found new homes in the Tesla Memorial Society of New York, the Tesla Engine Builders Association and the Tesla Universe website. Tesla and his inventions still fascinate fans of science fiction and computer games, and he has been celebrated in documentaries on his life and the *War of the Currents* on America's PBS and the BBC.

In 2003, Tesla Motors began producing electric vehicles in California, using engines based on Tesla's designs. They are now quoted on the NASDAQ stock exchange and have a European Headquarters at Maidenhead in the UK.

Tesla Magazine was launched on 10 July 2013. At the same time, the Tesla Science Foundation joined forces with Belgrade's Nikola Tesla Museum to create a travelling exhibition called *Tesla's Wonderful World of Electricity* which toured the United States and Canada. Its aim was to get Tesla the recognition he deserves in America. With new movies being made about his life, it is possible that it is not too late - this goal may still be achieved.

Left: *Prophet of Science* comic strip depicting Tesla's life.

Above: The Tesla Model S electric car, 2013.

Nikola Tesla reading in front of the spiral coil of a high-frequency transformer at his East Houston Street laboratory in New York.

SOURCES AND CITATIONS

This popular biography of Nikola Tesla has been produced by research carried out using the archives of one of the most renowned institutions in the world – the British Library in London. Primary source knowledge from the publications of the Tesla Museum in Belgrade, Tesla's own autobiography, *My Inventions,* published in 1919, articles from contemporary electrical scientific journals and newspapers of the time, along with copies of Tesla's original patent diagrams, have all been blended with material from an extensive variety of previously published books on Tesla.

The resulting recreation of the life and times of Nikola Tesla, is an informative and entertaining introduction to the great man. There are other more academic publications available should the reader wish to delve more deeply.

Books that were especially useful in the preparation of this book are listed below along with the abbreviations shown in square brackets to refer to them in the full list of citations that follows.

SOURCES

Burgan, Michael, *Nikola Tesla: Physicist, Inventor, Electrical Engineer,* (Minneapolis, Compass Point Books, 2009) [Burgan, *NT: PIEE*]

Carlson, W. Bernard, *Tesla: Inventor of the Electrical Age* (Princeton, NJ: Princeton University Press, 2013) [Carlson, *Tesla*]

Cheney, Margaret, *Tesla: Man Out of Time* (Englewood Cliffs, New Jersey: Prentice-Hall, 1986) [Cheney, *Man Out of Time*]

Cheney, Margaret & Uth, Robert, *Tesla, Master of Lightning* (New York: Barnes & Noble, 1999) [Cheney & Uth, *Master of Lightning*]

Jonnes, Jill, *Empires of Light: Edison, Tesla, Westinghouse, and the Race to Electrify the World* (New York: Random House, 2003) [Jonnes, *Empires of Light*]

Martin, Thomas Commerford, *Inventions, Researches and Writings of Nikola Tesla.* (First published 1894, reprinted New York: Barnes &Noble, 1995) [TCM, *IRWNT*]

Mrkich, Dan, *Nikola Tesla: The European Years.* (Ottawa: Commoners Publishing, 2004) [Mrkich, *Tesla: The European Years*]

O'Neill, John J., *Prodigal Genius: The Life of Nikola Tesla* (First published 1943, reprinted London: Granada Publishing, 1968) [O'Neill, *Prodigal Genius*]

Passer, Harold, *The Electrical Manufacturers, 1875-1900* (Cambridge, Mass: Harvard University Press, 1953) [Passer, *The Electrical Manufacturers*]

Ratzlaff, John T., *Tesla Said* (Millbrae, Tesla Book Co, 1984) [*Tesla Said*]

Seifer, Marc J. *Wizard: The Life and Times of Nikola Tesla, Biography of a Genius* (Secaucus, New Jersey: Birch Lane Press, 1996) [Seifer, *Wizard*]

Stewart, Daniel Blair, *Tesla: The Modern Sorcerer* (Berkeley, California: Frog Ltd., 1999) [Stewart, *Modern Sorcerer*]

Tate, Alfred O., *Edison's Open Door: The Life Story of Thomas A. Edison, A Great Individualist* (New York: E.P. Dutton, 1938) [Tate, *Edison's Open Door*]

Tesla, Nikola & Childress, David H., *The Tesla Papers* (Stelle, Illinois: Adventures Unlimited, 2000) [NT & Childress, *The Tesla Papers*]

Tesla, Nikola, *My Inventions and Other Writings* (1919) www.teslasautobiography.com [NT, *My Inventions*]

Tesla, Nikola *Lectures, Patents, Articles (1856-1943).* (Belgrade: Tesla Museum, 1956) [NT, *Lectures, Patents, Articles*]

Valone, Thomas, *Harnessing the Wheelwork of Nature: Tesla's Science of Energy,* 21-23 (Kempton, Ill: Adventures Unlimited Press, 2002) [Valone, *Wheelwork of Nature*]

CITATIONS

INTRODUCTION

1. *Electrical Experimenter,* January 1919, 614
2. NT, *My Inventions*
3. *New York Times,* January 23, 1943, 24

CHAPTER 1

1. NT, *My Inventions*
2. *New York World,* July 22, 1894
3. Mrkich, *Tesla: The European Years,* 55
4. Seifer, *Wizard,* 5
5. Burgan, *NT: PIEE,* 20
6. *Tesla Said,* 283-84
7 – 19. NT, *My Inventions*
20. Seifer, *Wizard,* 13 (citing NT, April 20, 1893)

21. Cheney & Uth, *Master of Lightning*, 9

22 – 26. NT, *My Inventions*

CHAPTER 2

1. *Electrical Engineer*, September 24, 1890

2. Carlson, *Tesla*, 26

3 – 7. NT, *My Inventions*

8. Mrkich, *Tesla: The European Years*, 10-11

9. Seifer, *Wizard*, 17

10. Seifer, *Wizard*, 18

11. NT, *My Inventions*

12. *Detroit Free Press*, August 10, 1924

13. Mrkich, *Tesla: The European Years*, 76

14 – 16. NT, *My Inventions*

17. NT, *Lectures, Patents, Articles*, 173

18. Mrkich, *Tesla: The European Years*, 100

19 –21. NT, *My Inventions*

22. *Electrical World*, May 25, 1889

23. Seifer, *Wizard*, 29

24. NT, *My Inventions*

25. Cheney, *Man Out of Time*, 25

26 – 28. NT, *My Inventions*

29. Carlson, *Tesla*, 68 (citing Tesla in *Complaint's Record on Final Hearing*, Volume 1: *Westinghouse vs. Mutual Life Insurance Company and H.C. Mandeville* [1903]; NT, "Electric Magnetic Motor", US Patent 424,036 [filed May 20,1889, granted March 25,1890]; TCM, *IRWNT*, 69)

30. NT, *My Inventions*

31. Hunt, Samantha, *The Invention of Everything Else*, 52 (Houghton Mifflin, 2008)

32. *The Essential Tesla*, 135 (Radford VA: Wilder Publications, 2007)

33. *New York Sun*, July 12, 1937

CHAPTER 3

1. *Tesla Said*, 208

2. Stewart, *Modern Sorcerer*, 163

3. *Buffalo New York News*, August 30, 1896

4 – 6. NT, *My Inventions*

7. *Century*, February 1894

8 – 9. NT, *My Inventions*

10. Tate, *Edison's Open Door,* 149

11. Carlson, *Tesla*, 71

12 – 13. NT, *My Inventions*

14. Cheney & Uth, *Master of Lightning*, 20

15. Thulesius, Olav, *Edison in Florida: The Green Laboratory*, 120. (Gainesville: University Press of Florida, 1997)

16. Seifer, *Wizard*, 40-41 (citing Tesla Electric Co., [advertisement], *Electrical Review*, September 14, 1886)

17. NT, *My Inventions*

18. *Tesla Said*, 280

19. O'Neill, *Prodigal Genius*, 65

20. Seifer, *Wizard*, 42 (citing *Electrician and Electrical Engineer*, 1886)

21. TCM, *IRWNT* 111-13, 429-31

22. *Electrical Experimenter*, March 19, 1919

23. Carlson, *Tesla*, 81 (citing Tesla in *Complaint's Record on Final Hearing*, Volume 1: *Westinghouse vs. Mutual Life Insurance Company and H.C. Mandeville* [1903]; Szigeti, 1889 deposition, A398)

24. Carlson, *Tesla*, 84-85 (citing NT, 78. 21 "Defendant's Brief, Derivation Electric Motor", in *Westinghouse Electric and Manufacturing Company vs. Dayton Fan and Motor Company*, 1900)

25. NT, *My Inventions*

CHAPTER 4

1. *New York Sun*, July 12, 1937

2. Carlson, *Tesla*, 90; also Klein, Maury, *The Power Makers; Steam, Electricity, and the Men Who Invented Modern America*; Davis, L.J. *Fleet Fire: Thomas Edison and the Pioneers of the Electric Revolution; Material History Review*, Vol. 33-37, 25

3. *The Essential Tesla*, 10 (Radford VA: Wilder Publications, 2007)

4. Seifer, *Wizard*, 50 (citing Westinghouse, internal memorandum, July 5, 1888; Passer, *The Electrical Manufacturers*)

5. Passer, *The Electrical Manufacturers*, 279

6. Carlson, *Tesla*, 113 (citing Tesla in *Complaint's Record on Final Hearing*, Volume 1: *Westinghouse vs. Mutual Life Insurance Company and H.C. Mandeville* [1903])

7. *Electrical World*, March 21, 1914

8. *New York Times*, July 31, 1888

9. Moran, Richard, *Executioner's Current, Thomas Edison, George Westinghouse, and the Invention of the Electric Chair* (New York: Vintage Books, 2003)

10. *Electrical World*, September 20, 1924

11 – 13. Seifer, *Wizard*, 54 (citing Scott to Tesla, July 10, 1931)

14. Cheney & Uth, *Master of Lightning*, 24

15. McPherson, Stephanie Sammartino, *War of the Currents: Thomas Edison vs Nikola Tesla*, 40 (Breckenridge, Co.: 21st Century Books, 2012)

16. *Electrical Review*, June 30, 1888

17. Munson, Richard, *From Edison to Enron*, 26-27 (Praeger, 2005)

18. Brandon, Craig, *The Electric Chair: An Unnatural American History*, 82 (McFarland & Co, 2009)

19. *New York Sun*, August 25, 1889

20. Jonnes, *Empires of Light*

21. *Electrical Review*, August 16, 1890

22. *New York Times*, August 6, 1890

23. O'Neill, *Prodigal Genius*, 83

24. Lamme, B. *Benjamin Garver Lamme: An Autobiography*, 60. (New York: G.P. Putnam, 1926)

25. *Electrical Review of London*, August 14, 1896

26. O'Neill, *Prodigal Genius*, 77

27. *Electrical Engineering*, May 1934

28. *Electrical World*, April 18, 1891

29. *Harper's Weekly*, July 1891

30. *Electrical World*, March 26, 1892

31. *The Essential Tesla*, 63. (Radford,VA: Wilder Publications, 2007)

32. Cheney, *Man Out of Time*, 293

33. NT, *My Inventions*

34. *Electrical Review*, March 19, 1892

35. Seifer, *Wizard*, 415 (citing Tesla to JP Morgan Jr, November 21, 1924)

CHAPTER 5

1. Belohlavek, Peter & Wagner, John W., *Innovation: The Lessons of Nikola Tesla*, 88 (Blue Eagle Group, 2008)

2. *Tesla Journal*, November 2 & 3, 1982-83, 25

3. Seifer, *Wizard*, 99; Carlson, *Tesla*, 193

4. *Electrical Experimenter*, May 1919

5. NT, *My Inventions*

6. *New York World*, April 13, 1930

7. TCM, *IRWNT*, 318-20

8. *Electrical Engineer*, June 28, 1893
9. Anderson, Leland (ed) *Nikola Tesla on His Work with Alternating Currents*, p XV (Breckenridge, Co.: 21st Century Books, 2002)
10. *Century*, April 1895
11. Carlson, *Tesla*, 161 (citing Tesla to Westinghouse, September 12, 1892)
12. TCM, *IRWNT*, 319-20
13. Cameron, William, *The World's Fair*, 318 (New Haven, Conn.: James Brennan & Co, 1894)
14. Barrett, J.P. *Electricity at the Columbian Exposition*, 168-69 (Chicago: R.R. Donnelly, 1894)
15. *Chicago Tribune*, August 26, 1893
16. *Electrical Experimenter*, March 1919
17. *World*, July 22, 1894
18. *Review of Reviews*, March 1894
19. O'Neill, *Prodigal Genius*, 329
20. Johnson, Robert Underwood *Remembered Yesterdays*, 400 (first published in *Century Illustrated Monthly Magazine*, 1923, reprinted Literary Licensing, LLC, 2013)
21. Seifer, *Wizard*, 128
22. Cheney & Uth, *Master of Lightning*, 52
23. O'Neill, *Prodigal Genius*, 324
24. *New York World*, July 22, 1894
25. *English Mechanic and World of Science*, March 8, 1907
26. *Letters of Swami Vivekananda*, 281-83

CHAPTER 6

1. NT, *My Inventions*
2. Adams, Edward Dean, *Niagara Power; History of the Niagara Falls Power Company 1886-1918*
3. *New York Times*, July 16, 1895
4. Cheney & Uth, *Master of Lightning*, 61
5. *Tribute to Nikola Tesla, 1961*, LS-41
6 – 8. *Electrical World*, April 4, 1896
9. *New York World*, July 22, 1894
10. *Troy Press*, April 20, 1895
11 – 12. Seifer, *Wizard*, 149
13. *Philadelphia Press*, June 24, 1895
14. *Electrical Experimenter*, March 1917
15. Seifer, *Wizard*, 158 (citing Comfort, May 1896, and untitled newspaper clipping, Omaha, Nebraska, June 14, 1896)
16. *New York World*, March 8, 1896

17 – 18. *Electrical Review*, March 11, 1896
19. Bechol, Keith, *The Electrifying Nikola Tesla*, 15; *Electrical Review*, April 8, 1896
20. *Electrical Review*, December 1, 1896
21. *New York Evening Journal*, December 2, 1896
22. *New York Morning Journal*, August 10, 1897
23. *Niagara Falls Gazette*, July 20, 1896
24. *Buffalo Evening News*, January 12, 1897
25. *Tesla Journal*, 6 & 7, 1989-90, 4-10
26. De Forest, Lee, *Father of Radio: An Autobiography*, 76, 81, 85 (Willcox & Follet, 1950)

CHAPTER 7

1. *Scribners Monthly*, 1897, 527-28
2. *Tribute*, 25-27
3. Seifer, *Wizard*, 189
4. *New York Times*, April 7, 1897
5. *Scribners Monthly*, 1897, 527
6. Seifer, *Wizard*, 194
7. *Brooklyn Eagle*, July 11, 1935
8 – 9. Benson, Allan, *The World Today*, 1915, 173
10. NT, *My Inventions*
11. *New York Times*, May 3, 1898
12. *Current Literature*, February 1899, 136-37
13 – 14. *Electrical Engineer*, November 17, 1898
15. *Century*, June 1900
16. *Public Opinion*, December 1, 1898
17 – 18. *Electrical Engineer*, November 24, 1898
19. Seifer, *Nikola Tesla: Psychohistory*, 272 (citing survey by Ratzlaff & Anderson, 1979)
20. *Electrical Engineer*, November 24, 1898
21. *Electrical Review*, January 5, 1898
22 – 23. *Century*, June 1900
24 – 25. NT, *Lectures, Patents, Articles*, A-122

CHAPTER 8

1. *Mountain Sunshine*, July-August 1899
2. *Colorado Springs Evening Telegraph*, May 17, 1899
3. *Colorado Springs Evening Telegraph*, June 2, 1899

4. Carlson, *Tesla*, 267-68 (citing Hull, Richard L. *The Tesla Coil Builder's Guide to the Colorado Springs Notes of Nikola Tesla*, A28; Cheney & Uth, *Master of Lightning*, 87; Tesla vs Fessenden, 24; *New York Sun*, January 3, 1901)
5. Seifer, *Wizard*, 218 (citing Tesla to Astor, September 10, 1900)
6. Carlson, *Tesla*, 271 (citing *Tesla, Research Notes, Colorado Springs, 1899-1900*, 69)
7. Carlson, *Tesla*, 272 (citing Seifer, *Wizard*, 471)
8 – 9. *Electrical World and Engineer*, March 5, 1904
10 – 11. *Collier's Weekly*, February 19, 1901
12. Seifer, *Wizard*, 220 (citing *Pyramid Guide*, 1977)
13. *New York Sun*, January 3, 1901
14. Seifer, *Wizard*, 224
15. www.teslasociety.com/mars
16. Hawthorne, Julian, *Philadelphia North American*, 1901
17. Tesla, speech accepting the Edison Medal, May 18, 1917
18. Carlson, *Tesla*, 282 (citing Tesla to Johnson, October 1, 1899)
19. Tesla, *Colorado Springs Notes*, October 23, 1899; Hull, *Coil Builder's Guide*, 90
20. Seifer, *Wizard*, 227 (citing Tesla to U.S. Navy, September 27, 1899)
21. Howeth, L.S. *History of Communications-Electronics in the United States*, ch4. (Washington: United States Government Printing Office, 1963)
22. *Electrical Experimenter*, October 1919
23. Seifer, *Wizard*, 233-34 (citing *Electrical Experimenter*, October 1919)
24. O'Neill, *Prodigal Genius*, 186-87
25. *Colorado Springs Gazette*, October 30, 1903
26. Cheney & Uth, *Master of Lightning*, 87
27. O'Neill, *Prodigal Genius*, 193-94

CHAPTER 9

1. Seifer, *Wizard*, 236 (citing *Politika*, April 27, 1927)
2. *Tesla Journal*, 26-29
3. *New York Times*, March 20, 1907

4. *Electrical World & Engineer*, January 7, 1905

5. *Popular Science Monthly*, July 1900

6. *Colliers*, February 9, 1901

7. *Everyday Science and Mechanics*, December 1931

8. Seifer, *Wizard*, 246

9. *New York Herald Tribune*, September 27, 1927

8. Seifer, *Wizard*, 246

10. Cheney, *Man Out of Time*, 158; O'Neill, *Prodigal Genius*, 198

11. Seifer, *Wizard*, 255

12. Seifer, *Wizard*, 260

13. *Long Island Democrat*, August 27, 1901

14. *Electrical Review*, January 12, 1901

15. Seifer, *Wizard*, 267

16. *New York World*, April 13, 1930

17. Cheney, *Man Out of Time*, 203

18 –19. *New York Times*, January 14, 1902

20. Carlson, *Tesla*, 338 (citing Tesla to Morgan, January 9, 1902)

21. Carlson, *Tesla*, 339 (citing Tesla to Morgan, January 9, 1902)

22. Cheney & Uth, *Master of Lightning*, 100

23. *Port Jefferson Echo*, March 25, 1902

24. Seifer, *Wizard*, 297 (citing Tesla to Morgan April 22, 1903)

25. *Electrical World and Engineer*, March 5, 1904

26. Seifer, *Wizard*, 322-23

27. *New York Times*, February 14, 1909

28. Bergstresser, Ralph, *Dr. Nikola Tesla: The Forgotten Super Man of Our Industrial Age*, 6-7 (Mokelumne Hill Press, 1996)

29. *New York Times*, October 16, 1907

30. *Wireless Telegraphy and Telephony*, 1908

CHAPTER 10

1. Carlson, *Tesla*, 366 (citing Tesla to Viereck, December 17, 1934)

2. *New York World*, January 6, 1908

3. *New York Times*, June 8, 1908

4. *New York Herald*, October 15, 1911

5. *The Bee*, Danville, Virginia, March 2, 1928

6. *New York Times*, May 23, 1909

7. *New York Herald*, October 15, 1911

8. Stewart, *Modern Sorcerer*, 399

9. Paijmans, Theo, *Free Energy Pioneer: John Worrell Keely*, 139 (Adventures Unlimited Press, 2004)

10 –11. *New York Herald Tribune*, October 15, 1911

12. Carlson, *Tesla*, 373 (citing Seifer, *Wizard*, 362-66)

13. *National Press Reporter*, May 1912

14. O'Neill, *Prodigal Genius*, 173

15. Seifer, *Wizard*, 47-49 (citing correspondence)

16. NT & Childress, *The Tesla Papers*, 120

17. *Electrical Review and Western Electrician*, May 20, 1911

18 –20. *New York Times*, May 16, 1911

21. Seifer, *Wizard*, 352-53 (citing correspondence)

22. Seifer, *Wizard*, 354–55

23 – 25. *New York Times*, August 18, 1912

26. Seifer, *Wizard*, 358-59

27. *New York Sun*, March 24, 1912

28. *Electrical World*, June 28, 1912

29. *New York Times*, July 21, 1912

30. Seifer, *Wizard*, 360 (citing Tesla to Morgan Jr, March 19, 1914)

31. Seifer, *Wizard*, 370 (citing correspondence)

32. Seifer, *Wizard*, 370 (citing Hammond Collection, National Archives)

33. *New York Times*, September 6, 1914

34. *New York Times*, April 23, 1915

35. NT & Childress, *The Tesla Papers*, 138

36. Seifer, *Wizard*, 371 (citing Ratzlaff & Anderson, 100)

37. Jolly, W. P., *Marconi*, 225 (Stein and Day 1972)

38. *Los Angeles Examiner*, May 13, 1915

39. NT & Childress, *The Tesla Papers*, 136

40. *New York Times*, October 6, 1912

41. Tesla, *Nikola Tesla on His Work with Alternating Currents*, 105

42. *New York Times*, July 10, 1915

43. Seifer, *Wizard*, 377 (citing Lloyd Scott, Naval Consulting Board of the U.S., Washington: Government Printing Office, 1920)

CHAPTER 11

1. *New York Times*, November 7, 1915

2 – 3. *New York Times*, October 3, 1915

4. Carlson, *Tesla*, 375 (citing Seifer, *Wizard*, 380)

5. Seifer, *Wizard*, 378 (citing Johnson to Tesla, March 1916)

6. O'Neill, *Prodigal Genius*, 229

7. Seifer, *Wizard*, 379 (citing O'Neill, *Prodigal Genius*, 229)

8 – 9. *New York Times*, December 8, 1915

10. *Electrical Experimenter*, August 1917; also *Century*, June 1900

11. Cheney, *Man Out of Time*, 213

12. O'Neill, *Prodigal Genius*, 229-30

13 – 15. O'Neill, *Prodigal Genius*, 231

16. O'Neill, *Prodigal Genius*, 235

17 – 19. *Tesla Said*, 179-80 (citing Minutes of the Annual Meeting of the American Institute of Electrical Engineers, Held at the Engineering Societies Building, New York City, Friday Evening, May 18, 1917

20. *Tesla Said*, 181 (citing Minutes of the Annual Meeting of the American Institute of Electrical Engineers, Held at the Engineering Societies Building, New York City, Friday Evening, May 18, 1917

21. *Tesla Said*, 181-84 (citing Minutes of the Annual Meeting of the American Institute of Electrical Engineers, Held at the Engineering Societies Building, New York City, Friday Evening, May 18, 1917)

CHAPTER 12

1. Seifer, *Wizard*, 387 (citing Tesla to Waldorf-Astoria, July 12, 1917)

2. *New York Times*, December 8, 1915

3. Seifer, *Wizard*, 386 (citing J.B. Smiley, J.B., to Frank Hutchins, July 16, 1917)

4. Seifer, *Wizard*, 392-93 (citing Tesla to George Scherff, July 26, 1917)

5. Cheney, *Man Out of Time*, 174

6. *Electrical Experimenter*, September 1917

7. *New York Sun*, August 5, 1917

8. NT, *My Inventions*

9. Seifer, *Wizard*, 92-94 (citing correspondence)

10. Cheney, *Man Out of Time*, 223 (citing NT, *My Inventions*)

11. *Electrical Experimenter*, February 1919

12. Seifer, *Wizard*, 405 (citing Tesla to Gernsback, November 30, 1921)

13. Seifer, *Wizard*, 396-97 (citing Tesla to Morgan Jr, June 13, 1917)

14. Seifer, *Wizard*, 397 (citing Tesla to Scherff, September 25, 1917)

15. Seifer, *Wizard*, 397-98 (citing Tesla to Scherff, December 25, 1917)

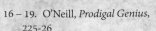

16 – 19. O'Neill, *Prodigal Genius*, 225-26

20. Seifer, *Wizard*, 402-03

21. *International Science and Technology*, November 1963

22. *New York Times*, January 19, 1919

23. *New York Times*, February 3, 1919

24. Cerf, C. & Navasky, V., *The Experts Speak*, 259 (New York: Villard, 1998)

CHAPTER 13

1. O'Neill, *Prodigal Genius*, 316

2 – 4. *Electrical Experimenter*, February 1919

5 – 7. O'Neill, *Prodigal Genius*, 291

8. O'Neill, *Prodigal Genius*, 291-92

9 – 10. *New York Times*, June 22, 1932

11 – 13. O'Neill, *Prodigal Genius*, 292

14 – 15. O'Neill, *Prodigal Genius*, 293-294

16. O'Neill, *Prodigal Genius*, 290

17. O'Neill, *Prodigal Genius*, 292-93

18. O'Neill, *Prodigal Genius*, 289-290

19 – 20. O'Neill, *Prodigal Genius*, 293

21 – 26. Seifer, *Wizard*, 406-08 (citing Nikola Tesla v. George Bold Jr, Suffolk County Supreme Court, April 1921)

27. Adrian Potter, *FBI report of Friends of Soviet Russia*, 1921-23

28. *New York World*, November 21, 1926

29 – 30. *Politika*, July 8, 1956

31. *Colliers*, January 30, 1926

32. O'Neill, *Prodigal Genius*, 289

33 –35. *New York World*, November 21, 1926

36. Cheney, *Man Out of Time*, 283

CHAPTER 14

1. Burgan, *NT: PIEE*, 87

2. Seifer, *Wizard*, 417

3. Valone, *Wheelwork of Nature*, 21-23

4. Seifer, *Wizard*, 417 (citing Flowers, John B., "Nikola Tesla's Wireless Power System and Its Application to the Propulsion of Airplanes", August 8, 1925)

5. *Telegraph & Telephone Age*, October 16, 1927

6. *Proceedings of the 1988 Tesla Symposium*, Colorado Springs, 8-11

7. NT & Childress, *The Tesla Papers*, 88

8. *Philadelphia Public Ledger*, November 2, 1933

9 – 10. *Time*, July 20, 1931

11 – 15. *New York Times*, July 5, 1931

16 – 17. *Time*, July 20, 1931

18 – 23. *New York Times*, July 11, 1931

24 – 26. *Kansas City Journal-Post*, September 10, 1933

27. *Charleston Daily Mail*, September 10, 1933

28. *Everyday Week Magazine*, October 21, 1934

29. *New York Times*, July 11, 1934

30. tfcbooks.com/teslafaq/q&a_011 (citing Tesla to Morgan Jr, November 29, 1934)

31. *New York Times*, July 11, 1934

32 – 35. *Baltimore Sun*, July 12, 1940

36. *New York Herald Tribune*, July 11, 1934

37. tfcbooks.com/tesla/1935-00-00

38. *Newsweek*, May 19, 1986

39. Seifer, *Wizard*, 428 (citing Tesla to Laemmle, July 15, 1937; Gabler, Neal, *An Empire of Their Own*, 58, Crown Publishing, 1988)

40. Siegel, Mark, *Hugo Gernsback: Father of Modern Science Fiction* (Borgo Pr, 1988)

41. *New York Times*, May 1, 1937

42. *The 1984 Tesla Centennial Proceedings*, 50-58

43 – 48. *New York Times*, July 11, 1935

49. *New York Times*, February 6, 1932

50 – 51. *Brooklyn Eagle*, July 10, 1932

52. *New York Herald Tribune*, August 18, 1935

53. teslasociety.com/ntinn

CHAPTER 15

1. *San Antonio Sunday Light*, July 20, 193

2 – 3. *New York Times*, February 6, 1932

4 – 6. *New York Times*, July 11, 1937

7. Seifer, *Wizard*, 435

8 – 10. *Liberty*, February 9, 1935

11 – 13. Smithsonian.com, April 19, 2013

14. *New York Times*, March 21, 1962

15. Gertz, Elmer, *Odyssey of a Barbarian: The Biography of G.S. Viereck*, 24 (Buffalo, NY: Prometheus, 1978)

16. *Serbian Newsletter*, 1943

17. teslauniverse.com/nikola-tesla/timeline (citing King Peter's diaries, July 8, 1942)

18. O'Neill, *Prodigal Genius*, 272-73

19. Seifer, *Wizard*, 443 (citing O'Neill, *Prodigal Genius*)

20. Cheney, *Man Out of Time*, 265

21. pbs.org/tesla/ll

22. Seifer, *Wizard*, 446-48, 458; Carlson, *Tesla*, 390-91; Cheney, *Man Out of Time*, 265

23 – 25. *New York Times*, January 13, 1943

26. teslasociety.com/eulogy

27. *New York Times*, January 13, 1943

28. Seifer, *Wizard*, 445 (citing *New York Sun*, January 1943)

29. Seifer, *Wizard*, 444

30. scribd.com/doc/200969295/Columbia

31. Gernsback, Hugo, "NT: Father of Wireless, 1857-1943," *Radio Craft*, February 1943

32. *Serbian Newsletter*, 1943

33. Cheney, *Man Out of Time*, 268-89; Seifer, *Wizard*, 448-56

34. *Proceedings of the 26th Intersociety Energy Conversion Engineering Conference: IECEC-91*, August 4-9, 1991, Boston, Mass., Vol 2, 410

35. Cheney, *Man Out of Time*, 276

36. *Tesla Said*, 278

37. Cheney, *Man Out of Time*, 278; Seifer, *Wizard*, 446

38 – 39. Seifer, *Wizard*, 458-59 (citing E.E. Conroy to J. Edgar Hoover, FBI, October 17, 1945)

40. Cheney, *Man Out of Time*, 278-79

41. Seifer, *Wizard*, 260

42. Valone, *Wheelwork of Nature*, 133

43. *Aviation Week & Space Technology*, May 2, 1977

44. *Public Papers of the Presidents of the United States, Jimmy Carter*, 1977, 793

45. Carlson, *Tesla*, 394-95 (citing SAC, Cincinnati, to Director, FBI, August 18, 1983)

46. fas.org/spp/eprint/keegan

CHAPTER 16

1. Anderson, L. I., *John Stone Stone, Nikola Tesla's Priority in Radio and Continuous-Wave Radiofrequency Apparatus*, The AWA Review, Vol. 1, 1986

2. Carlson, *Tesla*, 397 (citing Wyn Wachhorst, *Thomas Alva Edison*; Steven Watts, *The People's Tycoon*)

INDEX

This edition published in 2014 by
Chartwell Books
an imprint of Book Sales
a division of Quarto Publishing Group USA Inc.
142 West 36th Street, 4th Floor
New York, New York 10018
USA

Reprinted in 2014 twice, 2015 twice
Revised reprint 2016

© 2014 Oxford Publishing Ventures Ltd

All rights reserved. No part of this publication may be reproduced, stored in a retrieval system, or transmitted in any form or by any means, without the prior permission of the publisher nor be otherwise circulated in any form of binding or cover other than that in which it is published and without similar condition including this condition being imposed on the subsequent purchaser.

Every effort has been made to ensure that credits accurately comply with information supplied. However, the publisher will be glad to rectify in future printings any inadvertent omissions brought to their attention.

ISBN-13: 978-0-7858-2944-7
ISBN-10: 0-7858-2944-X

Printed in China

Picture Credits

Cover: Front: © Photo Researchers / Alamy/ Back: © Pictorial Press Ltd / Alamy/ Tesla Patent 487796 1892 Internal: 4 © Photo Researchers / Alamy/ 15 © INTERFOTO / Alamy/ 16 ©INTERFOTO / Alamy/ 19 © INTERFOTO / Alamy/ 22 © RGB Ventures LLC dba SuperStock / Alamy/ 23 © Pictorial Press Ltd / Alamy/ 29 © Everett Collection Historical / Alamy/ 31 Tesla Patent 396,121/ 32 Tesla Patent 428,057/ 37 © Everett Collection Historical / Alamy/ 39 © North Wind Picture Archives / Alamy/ 43 Nikola Tesla Induction Motor Drawing/ 47 © Mary Evans Picture Library / Alamy/ 53 © MARKA / Alamy/ 57 © bilwissedition Ltd. & Co. KG / Alamy/ 58 Tesla Patent 1,119,732/ 70 © INTERFOTO / Alamy/ 78 Tesla_boat/ 78 Tesla Patent 613,809/ 82 Tesla_Colorado/ 95 Poldhu_radio_station/ 97 flickr_wardenclyffe/ 100 © Photos 12 / Alamy/ 101 © Pictorial Press Ltd / Alamy/ 105 Tesla Patent 1,061,206/ 111 © Hilary Morgan / Alamy/ 113 Electrical Experimenter 1919, Vol.22, p.124/ 116 © PF-(bygone2) / Alamy/ 117 © Archive Pics / Alamy/ 121 Tesla Society of New York/ 128 © Science Wonder Stories October,1929/ 129 © Frank R. Paul/ 138 Wardenclyffe demolition/ 146 © Mary Evans Picture Library / Alamy/ 151 © Time magazine 1931/ 154 Tesla_thought_camera/ 158 © Moviestore collection Ltd / Alamy/ 160 Science and Invention 1922, Vol.9, No.9, p.809/ 168 © PF-(bygone) / Alamy/ 172 © Popular Radio 1922, Vol.2, No.3, p.159/ 183 © ZUMA Press, Inc. / Alamy/ 184 © Pictorial Press Ltd / Alamy/ 186/187 © AF archive / Alamy